SETTING THE CONTEXT

Suppression of Enemy Air Defenses and Joint War Fighting in an Uncertain World

JAMES R. BRUNGESS, Lt Col, USAF

Research Fellow
Airpower Research Institute

Winner of the Air Force Historical Foundation's
1992 Colonel James Cannell Memorial Award

Air University Press
Maxwell Air Force Base, Alabama

June 1994

Library of Congress Cataloging-in-Publication Data

Brungess, James R.
Setting the context : suppression of enemy air defenses and joint war fighting in an uncertain world / James R. Brungess.
p. 242
Includes index.
1. Air interdiction—United States. 2. Unified operations (Military science)—United States. 3. World politics—1989–
I. Title.
UG703.B78 1994
355.4'22—dc20 94-19404
CIP

ISBN 1-58566-051-5

First Printing June 1994
Second Printing September 2004

Disclaimer

This publication was produced in the Department of Defense school environment in the interest of academic freedom and the advancement of national defense-related concepts. The views expressed in this publication are those of the author and do not reflect the official policy or position of the Department of Defense or the United States government.

This publication has been reviewed by security and policy review authorities and is cleared for public release.

Air University Press
131 West Shumacher Avenue
Maxwell AFB AL 36112-6615
http://aupress.maxwell.af.mil

Contents

Illustrations

Foreword

Suppression of enemy air defenses (SEAD) has long been a critical concern to advocates of air power. It is especially critical to Pacific Air Forces because air power offers a primary means of responding rapidly and effectively to areas of conflict within the Pacific arena.

Lt Col James R. Brungess was selected in February 1991 to conduct research into developing joint SEAD tactics that would be useful in the Pacific Command area of operations—specifically for the Far Eastern Military District of the Soviet Union. We deemed it an important subject, and Colonel Brungess had the necessary credentials to conduct the study. The world, however, changed dramatically while he was conducting his study. This altered both the character and direction of his research effort. In a few short months from March to December 1991, the United States entered a new era heralded by Desert Storm and the collapse of the Soviet Union. In the span of less than a year, the United States Air Force vindicated the time-honored doctrine and strategies of air power in Desert Storm and lost its most powerful adversary with the demise of the Soviet Union in December 1991. While the same weapons continue to exist, these events have irreversibly changed the context in which they will be used. Both events have not only changed the shape of the world; they have forever altered the way we look at air power and, more importantly, the way we must use air power.

Setting the Context: Suppression of Enemy Air Defenses and Joint War Fighting in an Uncertain World explains why SEAD has changed the basic fabric of air warfare. In discussing the familiar themes of the past and the emerging paradigms of the future, Colonel Brungess weaves a web of changing interrelationships among services, politico-military structures, and research and development strategies, as well as developing a novel methodology for assessing events critical to air power's future. It places SEAD within the context of a changed world by defining relationships, explaining evolutionary patterns, and establishing a criterion to measure success, as well as pointing the way for future acquisitions, research, and joint tactics development. It is a tour de force in acquiring the big picture of where we have been, where we are, and where we need to go.

This is an important book. It not only provides scholarly research on historical and pragmatic issues, it grapples with key issues affecting real warfighters fighting real wars—the kind of information that tacticians and strategists can use to save lives and win wars. You will find the book intellectually challenging and exhilarating; Colonel Brungess has provided new ways of looking at new phenomena as well as an iterative, logical way of relating SEAD to the past.

Without a doubt, the most important feature of the book is its emphasis on the continued importance of SEAD to air power. Colonel Brungess provides a compelling argument that the technological cat-and-mouse game between air attacker and ground defender will accelerate over the next several decades. This will occur, paradoxically, during times of peace, and not because of national policy but because of international market forces. The evolution of air defense technology is only beginning to share in the same technology that has made air power preeminent, and, in times of dwindling fiscal resources, this is something for the air warrior to worry about. We need to make the most of what we have. *Setting the Context* gives us some good ideas to get the most from our air power defense dollar in terms of electronic combat and SEAD.

I urge you to read this book carefully. It provides a unique and original view into the world of electronic combat and SEAD that will enrich as well as inform. To use the author's words, "For the brief while that the world may give us, now is the time to reflect on our purposes and the time to create visions of the future—a future we have a golden opportunity to influence, to build, and to protect."

Ron Iverson

RONALD W. IVERSON, Maj Gen, USAF
DCS Operations, Headquarters, PACAF

About the Author

Lt Col James R. Brungess

Lt Col James R. Brungess holds a master of arts degree in international relations from the University of Southern California, United Kingdom Program. He is a career electronic warfare officer (EWO) with extensive experience in tactical aviation. He has flown in all three tactical air forces—Tactical Air Command (TAC), United States Air Force Europe (USAFE), and Pacific Air Forces (PACAF)—and served on the staffs of two of them. Colonel Brungess began his EWO career as an AC-130A/H gunship "black crow" in Thailand. He was responsible for the defense of the AC-130 from radar-guided weapons. Colonel Brungess served from 1975 to 1986 as a squadron EWO for four separate F-111 and EF-111 squadrons, among numerous other responsibilities. In 1981 he was selected for the EF-111 initial cadre where he helped to develop EF-111 tactics and spearheaded major initiatives to integrate EF-111 aircraft into the general tactics schemes of other fighter aircraft. Since 1986, Colonel Brungess has served on the staffs of TAC and PACAF as a functional manager for the EF-111 and F-4G. In 1988 he was selected as chief of the electronic combat division, and he became deputy director of tactical operations for the Pacific Air Forces in 1990. Having completed Air War College (AWC) as a member of the class of 1992, Colonel Brungess formerly served on the AWC faculty as a defense economics instructor and currently serves as director of the AWC Professional Writing Program and chief of the Curriculum Support Branch.

Preface

This study is conceptual. Originally, it was to focus on the specific air defense suppression problems that Pacific Command (PACOM) would face over the next few years with specific attention being paid to North Korea and the Far Eastern military district of the Soviet Union. In December 1991 the Soviet Union disappeared. It no longer makes sense to focus on the specifics of joint suppression of enemy air defenses (JSEAD) strategies and tactics in PACOM when the entire world has changed. It makes much more sense to establish a context where new relationships between strategy and tactics—and between the evolutionary trends in technology and the uncertain international environment—can be illustrated as a first step in reclaiming a rational grip on a radically changed world.

No matter what the traditional underpinnings of suppression of enemy air defenses (SEAD) have been, the changed world has not diminished SEAD's increasing importance. If anything, SEAD's stock has increased dramatically. Air power has become the hallmark of executing national military power, and, consequently, SEAD finds itself as a precious commodity in a shrinking military larder.

The technological revolution of the late twentieth century seems to be gradually giving way to the information revolution—a revolution that will propel us into the twenty-first century. The speed of information interchange and the compelling power of international economic interdependence have reshaped the way we—and the rest of the world—can acquire military power, the way we look at "the threat," and the basic way air power relates to the changed world.

There is nothing new about suppressing air defenses as a strategic or tactical concept, but the changed world environment has radically altered the traditional relationships SEAD once had. SEAD still performs its classic roles, but because it is part of the information revolution itself, it performs them pervasively and quickly. Other functions simply have not had time to catch up.

More has changed than just titles and organizations. We find ourselves in the ironic position of trying to explain a reality using terms that probably no longer apply; more than anything else, we need measurement tools to make sense of what has happened. This study is an attempt to provide those tools and to seek a context that helps us understand our changed circumstances.

James R. Brungess

JAMES R. BRUNGESS, Lt Col, USAF
Research Fellow
Airpower Research Institute

Introduction

Defense suppression is nothing new. It has existed as a concept ever since warriors sought to destroy the enemy's ability to defend against attack. Suppression of enemy air defenses (SEAD) is the latest iteration of the defense suppression concept; its application to air power as a fundamental element in protecting friendly air attackers and destroying the enemy's ability to defend against air attack is the result of a long and natural evolutionary process. The evolution of war, however, has harnessed technology itself and accelerated SEAD's evolutionary process almost immeasurably. The impacts of technology, environment, available resources, national objectives, and time have influenced what SEAD is and—more specifically for this study—what it can and should become.

Air power, as a product of technology, has grown from oblivion to become a key element of power projection in 80 short years. SEAD, having grown along with it, has become one of air power's prime enablers. SEAD has been propelled by the same evolutionary forces that shaped air power, but it was the surprising growth of enemy air defense technologies and defensive tactics that gave SEAD its special impetus and its growing importance as an element in air power application. Just as air power has enabled immediate military force projection, SEAD has enabled air power by giving it untrammeled use of the skies over the battlefield. Denial of the SEAD tool—or even diminished emphasis—in modern war would be devastating to the nation's power projection capability.

The purpose of this study is to make sense of what is happening in the world US forces face now and will face in the future. The threat may be unknown, it may be diffuse and undefinable, but it exists. As long as war is a feature of the international geopolitical landscape, the likelihood of US involvement is very high. Characteristics of the modern world are its immediacy and the speed with which events move. This makes air power a natural instrument of national power in influencing events. But air power's effect is blunted significantly when adequate SEAD measures are not taken.

The end of the cold war presents SEAD practitioners with a dilemma. The traditional threat has diminished, but the real threat remains. The real threat is represented by the rapid growth of sophisticated air defense technologies worldwide. There is at the same time a tangible movement to divest US forces of SEAD technologies and assets required to suppress the modern—if diffuse and undefined—air defense threat. Figure 1 shows the dilemma clearly. In the top graph, the perception of the diminished threat leads to the logical conclusion that forces can be cut back. The bottom graph shows the real threat with respect to SEAD. While relative US

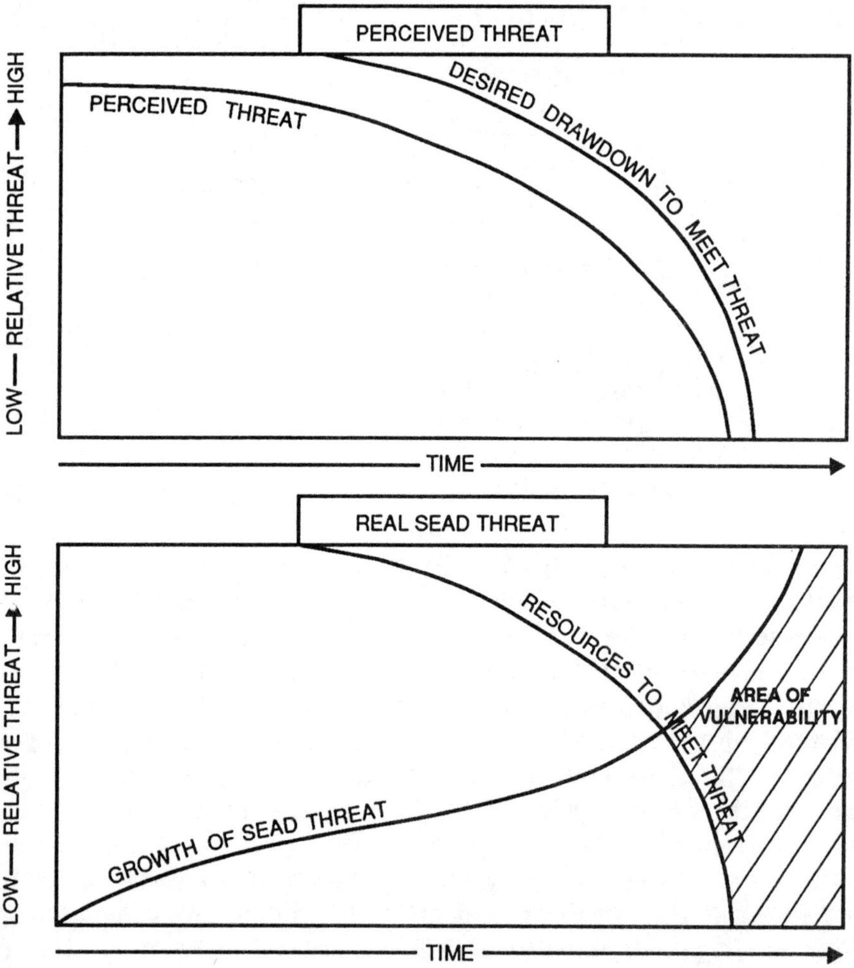

Figure 1. Real versus perceived SEAD threat.

SEAD superiority diminishes as a result of shrinking resources, those of the rest of the world increase. At some unknown time in the future—assuming the trend continues—US SEAD will be unable to defend US air power. Clearly, the SEAD community must prevent this from happening.

There are several themes that pervade this study. One of these is that SEAD is at the leading edge of air power's technological revolution. Another is that SEAD is expanding its traditional boundaries and is gradually infusing itself into the basic day-to-day tactics and strategies of air power practitioners. A third is that our basic way of looking at decision-making structures and measuring effectiveness with respect to SEAD needs to change to fit the real world; in short, we need to build a new frame of reference. A fourth theme is that SEAD forces a focus on objectives rather than authority, resulting in a natural bridge to conduct true "joint" SEAD applications. Greater joint cooperation among the services is not only mandated, it is desirable and necessary for effective future SEAD application on the battlefield.

There are six chapters in this study. Chapter 1 begins by taking a historical look at the foundations of SEAD, identifying pertinent variables, and tracing them through to the present to determine how they have changed.

Ten variables are traced through four historical benchmarks in the evolution of SEAD. The historical examination also explores the relationship between SEAD's growth and SEAD doctrine development.

Chapter 2 examines the development of meaningful criteria and measurements of effectiveness by which SEAD can be assessed. The chapter offers some suggestions on new criteria constructs—called "objective-based" criteria—that may prove useful in assessing the relative value of SEAD doctrine, strategy, tactics, and equipment. The examination details the evolution of approaches to SEAD criteria development by building four categories and following the growth and decline of each. The chapter also introduces a new tool for assessing change: the use of four specific continuums where the services' SEAD contributions can be placed with respect to each other. These continuums are distillations of the 10 variables in chapter 1 and form the basis of evaluation of both single-service and joint SEAD in subsequent chapters.

Chapter 3 examines how each service's culture has influenced the growth of SEAD and whether each service's use of SEAD has implications for joint SEAD application. Its primary themes are that each service applies SEAD according to its view of air power and that these views are converging. This knowledge is beneficial in two ways. First, by knowing how each service views the contribution of SEAD to its fighting doctrine, other services can employ their SEAD assets accordingly. Second, by each service knowing how SEAD fits into the other services' doctrinal frames of reference, all can gain a clearer view of what joint SEAD may look like.

Chapter 4 assesses the value of joint SEAD and shows how each of the services fits with respect to the continuums introduced in chapter 2. The principal themes of this chapter are that the movement toward joint SEAD is a natural outgrowth of converging service views on the use of air power and that JSEAD offers opportunities for enhanced air power effectiveness not available under single-service SEAD applications.

Chapter 5 is an analysis of the future threat and of how JSEAD strategies and tactics can be galvanized to defeat the threat. The key theme is that the integrated air defense systems (IADS) of the future will combine with enemy command, control, and communications (C^3) functions into one gigantic, cohesive net. The primary driver of this net is the ability to transfer information rapidly. Information is the lifeblood of the future threat; therefore, tactics and strategies aimed at information denial will cripple the future IADS net. The chapter goes on to suggest some specific JSEAD tactics based on information-denial strategies.

Chapter 6 is divided into three sections. The first section ties the themes together through a series of conclusions. The evolutionary theme predominates, but the chapter suggests that SEAD practitioners need to build an entirely new paradigm that accounts for the world as it really is, not as it was, or as SEAD practitioners wish it to be. This section also suggests that criteria development must focus on visionary objectives rather than short-term goals.

The second part of chapter 6 offers a series of recommendations that follow the general "train, equip, and organize" purpose statements of the peacetime military. These recommendations also suggest that education as well as training is required for SEAD practitioners and their superiors to understand the concepts that shape the modern IADS net and military environment and to retain competence in developing, building, and operating sophisticated SEAD weapons. The chapter recognizes that there is a bonding among strategy, tactics, and doctrine that seems to have gone full circle when viewed against the entire backdrop of military history. Maybe—as fundamental as the changes are in the world of SEAD—a larger framework is in operation that seeks equilibrium according to a set of variables that are imperceptible, perhaps unknowable.

If air power is a child of technology, then SEAD is surely a stepchild. Arguably, the most important technological advance in warfare—excepting aerospace and nuclear weapons—is the employment of the electromagnetic spectrum in the service of warfare. Just as aircraft reduced the field of combat from the context of distance and reframed it in terms of time, electronic combat has reframed the air power world of minutes and seconds into microseconds and nanoseconds.

SEAD's job in the future will be to cripple the information-based IADS net of a potential adversary *before* it has a chance to react and to keep it off balance until air power objectives can be met. The job will not be easy in light of political and economic challenges to the robust technology base needed to support the future effort. SEAD practitioners must state their case in terms of overall objectives and final outcomes rather than in technical jargon and short-term desires.

The stakes are high and getting higher. US military power faces its biggest challenge in a world where technology access continues to level the playing field. We will need every edge to maintain air superiority in future conflicts; having a potent, viable SEAD option may well be the determining factor.

Chapter 1

History and Doctrine

Embedded in the concept of suppression of enemy air defenses' (SEAD) evolutionary process is the historical view—the idea that SEAD began at an identifiable point in time and progressed through identifiable transitions. More pertinent is the idea that the process continues, that it has trends, and that it can be directed as a function of rational action.

The Evolutionary Construct

This chapter examines four points in time to establish milestones by which discrete comparisons of similar variables can be made. An appreciation of these milestones will reveal where SEAD fits in the larger framework of aerospace warfare and where it "needs to go" (directed evolution) to meet the larger objectives of military strategy in the context of defending the nation. The four periods of time chosen as benchmarks are: (1) the Vietnam War (1965–73), (2) Israeli operations in the Bekaa Valley (June 1982), (3) Operation El Dorado Canyon, Libya (April 1986), and (4) Operation Desert Storm, Iraq/Kuwait (January–March 1991).

In defining the time periods, we also must identify measurable variables to ascertain what changed and in what directions these changes impelled SEAD concepts. Because the time periods span almost 30 years of intense political, economic, and technological change, some of the variables will defy precise description. Nonetheless, they are precise enough to convey verifiable changes in direction and to portray a clearly discernible evolutionary process. These 10 variables are as follows.

The threat. Ultimately, SEAD is a contest between "us and them." By understanding and analyzing the enemy's ability to defend against air attack, we can configure tactics, define equipment requirements, and establish force and munitions levels. In knowing how the threat has changed, we can better understand why SEAD has reached its current preeminent status.

Definitional changes. The ways SEAD has come to be defined, both de facto and institutionally, show the direction and breadth of its growth.

Tactical application changes. The functional use of concepts and weapons at this basic level of modern combat is usually the first place pragmatic fixes to real problems become apparent. SEAD is no different on this account.

Organizational changes. Using a time-honored dictum from sociology—that viable constructs exist informally long before they become institutionalized—we will see how SEAD grew from necessary informal structure to institutional status and how it was composed as a function of the American war-fighting apparatus in its evolutionary process.

Force structure changes. This is a more specific subset of the organizational framework that delineates what types of equipment were purchased to solve the SEAD problem and how, over the 30-year period of time analyzed, the mix of aircraft, weapons, and tactics has altered the force structure with respect to SEAD.

Strategy changes. We will concentrate on general military strategy as much as possible, though it is in some respects impossible to ignore the effects of national strategy. SEAD has had enormous impact on air power strategy both from the perspective of building a tactical framework to execute it and as a concept with which to view a potential adversary.

Doctrinal changes. Defined as "*what we believe about the best way to conduct military affairs,*"[1] doctrine has been slow to change—glacially slow according to some. As doctrine is the defining enunciation of a large and conservative military institution, this is understandable. Nonetheless, what changes have been made reflect the general tenor of SEAD's evolutionary path.

Technological changes. The forces of technology have propelled many aspects of warfare, but the most important advances in air power have been in the uses of the electromagnetic spectrum and the harnessing of computer technology. An analysis of how each element of technology has contributed to the evolution of SEAD is beyond the scope of this study. The most telling evolutionary feature of technology, however, may be its unwitting (and often seductive) influences on weapons acquisition and fiscal resource allocation in the SEAD arena.

Political changes. These changes are both international and national. The will of the people as expressed through public opinion and governmental power structures has modified the ways in which military force can be applied and has affected the way SEAD has evolved in some unanticipated ways.

Economic changes. The general health and direction of the national and international economic milieu have had profound effects on the amount of capital available to pay for implements of war, and, consequently, the military has been forced to prioritize its shopping list. The effects this prioritization has had and still has on SEAD's evolutionary path is ironic. On the one hand, it limits the amounts and types of equipment that may be purchased. On the other hand, since the SEAD task must be accomplished to prosecute war to a favorable planner's conclusion, limiting resources has forced SEAD planners to become more creative in their approaches.

Isolating each of the above variables would be impossible as each is related in some way to the others. Numerous authors indicate that the relative influence of each of these variables on strategy, doctrine, operational art, and tactics changes dramatically from situation to situation.[2] Very few authors, however, delineate what forces determine the factors that dictate which variable will be preeminent in which situation—except with the obvious benefit of historical hindsight. Also, not all the factors pertain to the historical situations analyzed.

The matrix in figure 2, which is formed by combining historical benchmarks and pertinent variables, not only plots change over time but indicates the direction of each variable, defines trends, and shows how some basic relationships among the variables have changed over the last 30 years. The matrix can provide realistic options open to us in the reframing of SEAD concepts. The matrix maps some general concepts and examines the historical references in an attempt to substantiate the general trends presented. The matrix's *structure* is presented in figure 2.

	Vietnam	Bekaa Valley	Libya	Iraq/Kuwait
Threat				
Definitions				
Tactics				
Organizations				
Force Structures				
Strategy				
Doctrine				
Technology				
Political Base				
Economic Base				

Figure 2. Historical benchmark/variable matrix.

Vietnam

SEAD in Vietnam grew from a very simple fact: US forces were losing a large number of aircraft to various elements of the North Vietnamese air defense structure. There are three phases of the air war in Vietnam with which we will deal: Rolling Thunder, Linebacker I, and Linebacker II. Each was accompanied by a shift in military strategy and tactical focus.[3] Each phase also illustrated the growth of SEAD. Rolling Thunder had as its military objective an interdiction campaign to restrict the flow of supplies to the Vietcong inside South Vietnam. The national command authorities (NCA) hampered this effort by restricting American forces from attacking key targets and by failing to define a strategy in more precise military terms. The NCA's military objectives were unclear.[4] This period extended from December 1964 to the summer of 1968 and was characterized by piecemeal SEAD tactics. The defenses around the immediate target area were attacked *as a function of the target,* not as an element of the enemy's overall air defense structure.

As early as November 1964, American intelligence knew of Hanoi's efforts to build an air defense system based on "air deniability" wherein

> they used their total air defense system as an integrated entity to prevent the US Air Force, Navy, and Marine Corps from doing what they wanted to do in the skies over North Vietnam. Air deniability, as a strategy, was a lower, more basic form of warfare constituting, in its essence, a people's war in the air.[5]

The American tactical response to "air deniability" considerably lagged the intelligence estimate. American tacticians and strategists apparently did not see the need to develop an air campaign against the North Vietnamese integrated air defense system (IADS). In fact, they did not create specific SEAD tactics against any of the various elements of the North Vietnamese IADS until American air power began losing large numbers of aircraft to "the threat."[6] On 24 July 1965, an SA-2 brought down an F-4C. By the end of 1965, the North Vietnamese had constructed more than 60 surface-to-air missile (SAM) sites in the North and had brought down 25 of the 171 aircraft lost over the North that year. The North Vietnamese scored a hit for every 13 missiles fired.[7]

The definition of SEAD was tied directly to the immediate threat in order to protect fighter-bombers tasked against specific targets. The definition was defensive and reactive in nature. This definition, however, gradually expanded to include the development of a special mission—the "Wild Weasel" mission and the introduction of specially modified F-100s. The first Wild Weasels were equipped with rudimentary detection equipment to locate the radar-directed SAM and antiaircraft artillery (AAA) and to bomb the sites. This mission was extremely hazardous, and aircraft losses kept mounting.

The excessive loss of American aircraft to surface-to-air missile systems spurred the deployment of a special task force that delineated SAM vulnerabilities in 1966. The results were threefold: (1) the continued development of specialized aircraft and equipment dedicated to suppressing the radar-directed threat (SA-2, SA-3, various AAA radars) by detecting, identifying, and locating the emitter; (2) the continued use of antiradiation missiles (ARM) that used the emitters' electromagnetic signature as a homing point; and (3) a detailed analysis of the enemy systems' vulnerabilities, which could be used by fighter-bomber aircrews to defend themselves when under attack. For instance, the SA-2 Guideline missile was known to be slow-reacting in flight. If detected, certain aircraft maneuvers at the appropriate time would cause the missile to overshoot. Additionally, these early vintage systems were observed to have poor low-altitude capability (a range extending from ground level to about 1,500 feet).

Unfortunately, the tactical "cure" to this essentially medium altitude threat (nominally 1,500 feet) was to develop bombing tactics in the low-altitude regime. The reaction of pursuing low-altitude tactics in the target area appears peremptory in hindsight. As the medium-altitude regime was "denied" to them, aircrews avoided the threat by flying less than 1,500 feet above the ground. This "cure" placed American fighter aircraft in the heart of numerous AAA batteries' lethal envelopes. During Rolling Thunder, AAA accounted for approximately 85 percent of all aircraft losses.[8] The US "tactical response" that drove American attack aircraft to these daytime low-altitude operations appears, in retrospect, to have worked more to the advantage of the North Vietnamese's "air deniability" strategy than it did for air superiority as espoused by US air doctrine of the time.

After a series of modifications to various other aircraft proved undesirable for the specialized SEAD role, the Air Force chose the F-105G to perform the Wild Weasel mission. The F-105G was equipped with specialized electronic devices that allowed it to detect, identify, and locate SA-2 sites, some radar-directed AAA sites, and, later in the war, SA-3 sites. Armed with the AGM-45 Shrike ARM (a relatively short-ranged missile with a low single-shot probability of kill against the targeted emitter) and the longer-ranged, much more capable AGM-78 Standard ARM, the F-105G proved to be a very formidable system. Studies performed by the Air Force Electronic Warfare Center indicate, ironically, that the greatest suppression element the Wild Weasel possessed was psychological.[9] As much as 95 percent of its effectiveness may have been attributable to this phenomenon. When SAM operators suspected an F-105G to be in the area, they would simply turn off their radar rather than be a willing target. Although turning off the radar prevented the site's destruction, it also gave the attacking aircraft safe passage to and from their targets. No matter what positive effect this tactic had, the primary focus of the SEAD mission at this stage of the war was strictly as a function of the individual mission, not part of a concerted, coherent plan to destroy the enemy's air defense assets *prior* to mounting an offensive air campaign.

The narrow tactical focus used by tactics planners espousing piecemeal tactics directed at single sites—one at a time—is curious in that American intelligence knew how the North Vietnamese configured their IADS. US intelligence knew that these systems were the nascent manifestations of an integrated structure, not just random point-defense systems for high-priority targets. Intelligence analysts also suspected the reason for configuring the North Vietnamese air defense net in that way well before the bulk of US air power was committed to purposeful action. Oddly, US intelligence either failed to tell battlefield commanders the effects of such an IADS structure on attacking aircraft or battlefield commanders failed to appreciate the implications of such an enemy structure. The US failure to appreciate the overall strategic picture the North Vietnamese concept of air deniability represented was also a failure to adjust American doctrine and strategy to the realities of an enemy's changing technological environment. Even though US *tacticians* knew the enemy had made an effective move in the technological chess game, American *decision makers*, while apparently possessing knowledge of the overall picture, knowingly remained a move behind. The Korean doctrine, which drew a clear distinction between air superiority and interdiction missions—and did not even address SEAD per se—obscured the integrated nature of the threat, rudimentary though it may have been. The overall mental construct, which would have allowed a truer picture of events, did not exist during the early phases of the Vietnam War, and while there were visionaries among some tacticians, these visionaries were not yet among the generals.

The implications of air deniability were important to the definitional growth of SEAD. Gradually, US technological capability enabled aircraft to receive equipment that significantly altered SEAD tactics. These technological advances were represented by the introduction of the F-105G Wild Weasel, the AGM-45 Shrike, AGM-78 Standard ARM, the

EB-66, and the EA-6 radar jamming aircraft. These additions made it possible to develop a wide array of tactics ranging from the relatively unsophisticated Iron Hand tactic of using aircraft to find and bomb SAM sites to more elaborate "hunter-killer" defense suppression tactics that teamed fighter bombers and Wild Weasels.[10] With the new equipment and tactics, SEAD expanded to include attacking the enemy's radars using electromagnetic means (jamming) as well as destructive means (ARMs and iron bombs). However, strategists still considered SEAD as a one-on-one function—one SEAD asset for a specific target defense system—rather than a concept that sought to degrade the entire IADS apparatus. Not until the development of more formal organizational structures that specialize in the SEAD and electronic warfare (EW) functions will more encompassing tactics have an institutional voice with which to espouse these concepts.

Linebacker I (around April–October 1972) did little to alter the basic definitions even though tactics and equipment had undergone considerable evolution. Decision makers still labored with a conceptual framework mired in strategic dogma that lagged behind the tactical realities of air deniability.

During the technological chess game between the US and the Soviet-equipped North Vietnamese, a qualitative change took place in the North Vietnamese IADS that the US was slow to recognize. The strategy of air deniability in mid-1968 included 250 ground controlled interceptors (GCI), 1,500 radar-directed and optical AAA sites, and over 300 SAM sites netted together by a centrally controlled and directed integrated structure.[11] A sophisticated early-warning radar net consisting of the latest Soviet radars, communications apparatus, passive detection nets, and intelligence-gathering agencies fed an increasingly integrated air defense network. From the North Vietnamese view, the IADS was an indivisible organization composed of interdependent, interlocking parts comprising a complete "nervous system" and associated striking "muscle."[12]

By the end of Linebacker I, the North Vietnamese had reduced American air-to-air kill ratios from 2.2-to-1 to less than 2-to-1.5. During the summer and fall of 1968, there were periods when the North Vietnamese enjoyed a better than

1-to-1 kill ratio.[13] This was a direct result of an increased IADS efficiency and an apparent American inability to adjust tactics to meet the situation. On the positive side of the scale, only 14 aircraft losses between 10 May and 23 October 1968 (when Linebacker I ended) were due to SAMs while 27 were lost to enemy interceptors.[14] Wild Weasels and defense suppression raids against particular SAM sites were effective. The dilemma this situation posed—where one aspect of the enemy IADS was beaten down only to be supplanted by another—might better have been approached if SEAD had been directed at the entire IADS—including ground controlled interception—as opposed to an arbitrary separation of missions imposed by the "roles-and-missions" doctrine of the time.

Linebacker II (the 11-day air campaign over North Vietnam, 18–30 December 1972) introduced a new element into the already muddled arena: the use of Strategic Air Command (SAC) B-52s against the North Vietnamese IADS using SAC-only tactics. While the B-52 possessed a formidable EW capability against specific radar threats, it had never faced the wrath of the integrated North Vietnamese IADS of Hanoi and Haiphong. While Tactical Air Force (TAF) assets had amassed a wealth of knowledge on how to deal with the SA-2 threat, little of this knowledge was transferred to SAC aircrews. SAC carried the preponderance of the campaign. Unfortunately, the combination of poor tactics, a strong dose of overconfidence,[15] and a failure to integrate tactical SEAD resources with strategic assets resulted in the loss of 11 B-52s to SA-2s by the end of the fifth night of operations.[16] It was not until 26 December 1972 that air campaign planners made the decision to commence "an all-out attack on the North Vietnamese air defenses."[17] Once the IADS was defeated, it took three days to bring the North Vietnamese to the negotiating table. A curious footnote to the history of SEAD is that the first all-out SEAD campaign against an enemy IADS came at the very end of the Linebacker II air campaign as opposed to the beginning of it. The recognition that by making the enemy defenseless to air attack one was in a much better position to bargain and to fight came very late to those who waged air war against Vietnam. As clear a lesson as it appeared to be at the time, it was a lesson the Israelis would relearn in the Yom Kippur War of October 1973.

The extensive use of the electromagnetic spectrum as an aid to prosecuting war reached maturity with the Vietnam War. Electronic warfare—as waged by both the North Vietnamese and the Americans—was a harbinger of events to come. SEAD, as a concept, appears to have been a bonding element in tactics, strategy, and doctrine between electronic warfare and air superiority. In some ways, SEAD as practiced at the end of Linebacker II laid the groundwork for JSEAD in the post–Desert Storm military. Figure 3 shows how the war-fighting relationships among doctrine, strategy, and tactics were affected by SEAD.

SEAD also seems to have undergone parallel (and convergent) evolution in doctrine, strategy, and tactics. As a strategic element, it made sense. A concerted effort to destroy or deny the enemy's ability to detect, track, and fire on attacking aircraft would have a withering effect on the enemy's *ability* to pursue war aims as well as the enemy's *will* to continue. As a tactical element, the rapid suppression of enemy SAM emplacements and AAA sites using unpredictable patterns of

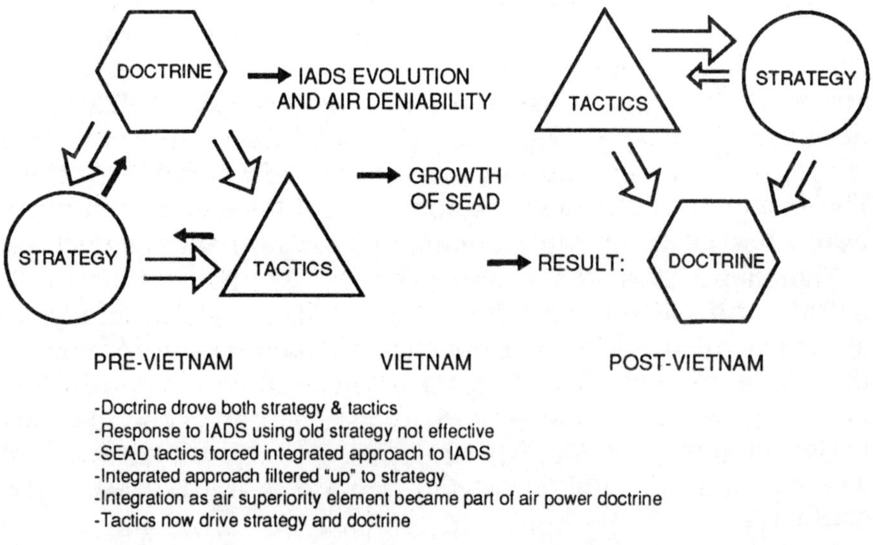

Figure 3. Doctrine/strategy/tactics relationships.

employment would prevent the enemy from ever seizing the tactical initiative. At the doctrinal level, a blend of SEAD strategies and air superiority tactics seems to offer an effective alternative to blunting an enemy's ability to wage war by neutralizing the enemy's means of protection from air attack.

SEAD patterns from Vietnam offer a pattern of convergence. As figure 3 shows, SEAD is a function of electronic warfare as well as a function of overall strategy and doctrine. Oddly, the conclusion of Linebacker II serves as an example of SEAD-driven doctrine. The tactical requirements of the war forced the use of piecemeal tactics that gradually worked their way into strategy and ultimately surfaced as part of the doctrinal lexicon. In short, SEAD pointed the way to how the campaign would be won not as a function of doctrine but as a "nuts-and-bolts" method of getting the job done (tactics).

Unlike electronic warfare and command, control, and communication countermeasures (C^3CM) which emphasize the technical aspects of manipulating the electromagnetic environment, SEAD's initial impetus stemmed from the fighter-bomber community's need to answer a specific defensive requirement. The recognition that SEAD served both "tactical" and "electronic warfare" interests may have served as a catalyst in bonding EW, C^3CM, and SEAD under the umbrella of electronic combat (EC). The enemy's use of the electromagnetic spectrum to track and shoot down friendly aircraft was a serious problem, and the requirement to neutralize it spurred vigorous activity in both the traditional technical EC areas and the pragmatic world of fighter tactics. The fusion of SEAD concerns to strategy also created a natural bridge to transmit strategic objectives to clear tactical goals.

Figure 4 shows the evolutionary relationship among EW, C^3CM, and SEAD. The evolutionary trend shows that the separate areas had come much closer together by the end of the Vietnam War. The Vietnam War was a vehicle for the evolution of SEAD mostly because it spanned such a long period of time. For that reason, we can compare what SEAD was at the beginning with what it became by the end of the conflict.

Table 1 is a matrix of SEAD's evolution during the Vietnam War. A number of patterns emerge from this presentation.

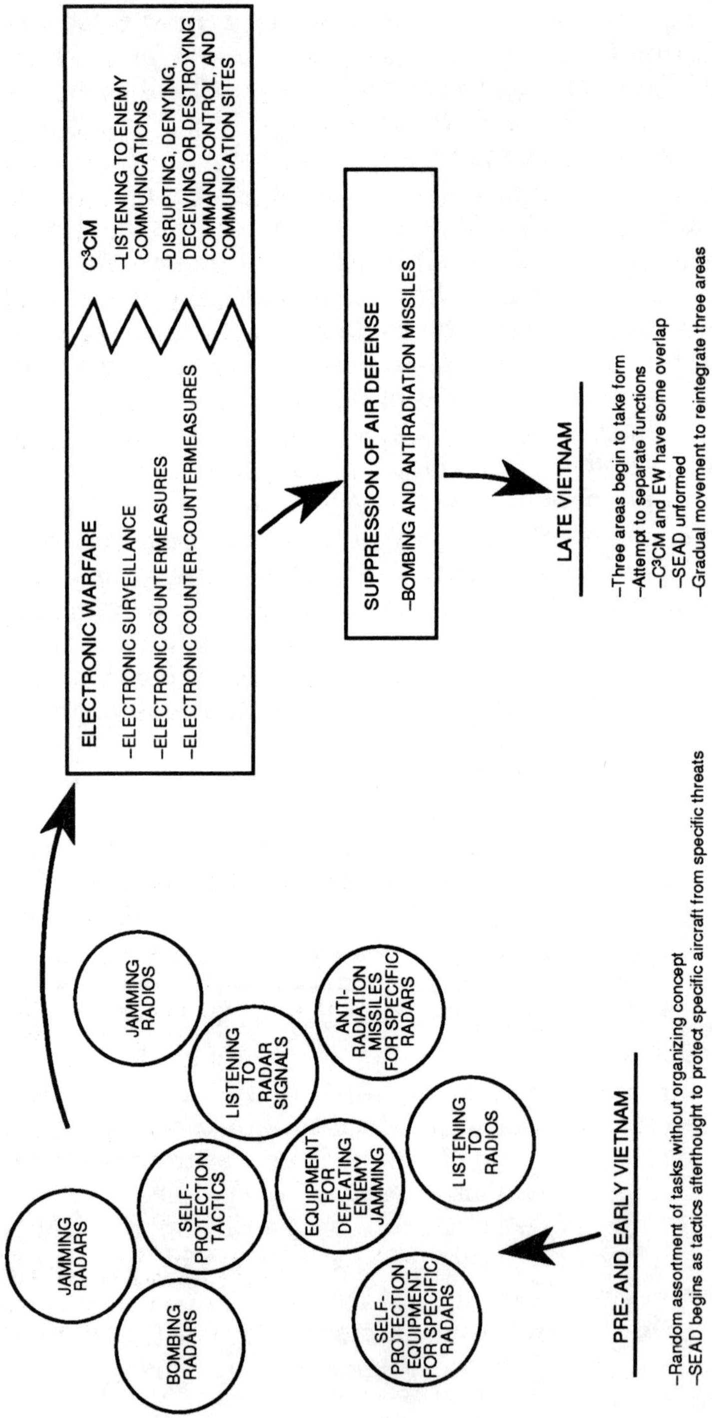

Figure 4. Evolution of Air Force SEAD relationships.

First is the concept that SEAD evolved from a one-on-one defensive tactic of protecting friendly aircraft prosecuting offensive strategies to a strategy aimed at destroying the enemy's defensive ability during an air attack. Second is the effect accelerating technology had on the cost of war and the integration of operations. Third is the tactical emphasis SEAD placed on seizing the initiative—that is, obligating the enemy to behave in predictable ways and taking advantage of the opportunities provided. Fourth is the fact that SEAD, because of its target-specific nature, obliged decision makers to define objectives in terms that were translatable to targets, times, and specific rules of engagement. Fifth is that in the debate between defensive and offensive precepts of waging war, SEAD brought back into balance a strategy that had been shifted toward unprotected use of air assets.

Table 1 summarizes the development of SEAD during the Vietnam War against each of the 10 variables given earlier in figure 2. Because the war was so long, reaction to technological innovation by Soviet and US weapons designers had a profound impact on SEAD. For this reason, the matrix portrays SEAD's development at the beginning and end of the conflict.

Table 1

Vietnam War Matrix
Air Deniability versus Air Superiority: 1965–73

	1965	1973
Threat	Limited SA-2 coverage (60 sites), growing AAA presence, beginnings of air interceptor force (less than 100). Naval presence limited to gun boats.	Fully integrated air defense system; introduction of SA-3, SA-7 radar-directed AAA; better trained pilots. More than 300 SAM sites, 1,500 AAA sites, 250 MiG-7, -19, -21.
Definitions	SEAD not defined as a role or mission.	Related to destruction of ground-based radar-directed air defense systems via antiradiation missiles. Clearly a support role.
Tactics	Reactive, not part of overall plan. Piecemeal; specific target and mission related.	Still reactive, but becoming more integrated. Innovations of Iron Hand and hunter-killer teams. SEAD campaign of 26 Dec 72 first concerted effort to target IADS.

Table 1 (cont'd)

	1965	1973
	Primarily defensive.	Stand-off weapons preferred. Jamming viewed as separate tactic.
Organization	None. SEAD, per se, did not exist except as a defensive consideration.	F-105Gs/EB-66s organized separately. No integration. Specialized planning cell developed to task both assets.
		Electronic surveillance measures treated separately.
		Overall command and control structure very complex. Chain of command and authority confusing, not clear-cut.
Force Structure	Air Force aircraft optimized for strategic warfare. F-105 not configured as maneuverable fighter. F-100, F-102, configured as strategic interceptors, not fighter bombers. Navy aircraft (F-4, A-4, A-6, A-7) configured for tactical warfare. Strategic aircraft (B-52) used in area interdiction.	Air Force adapts F-4, A-7 aircraft for tactical operations. Specialized aircraft modified from F-100s and F-105s. F-105Gs/EB-66s formed as separate squadrons.
Strategy	Nuclear deterrence. Defense suppression function of B-52 nuclear strike, self-contained. Carry over from Mitchell/Douhet, the impregnable bomber.	Strategic conventional war, using tactical and strategic assets interchangeably. SEAD integrated into destruction of N. Vietnamese infrastructure as function of overall plan.
	Initial Vietnam air strategy indeterminate. Could not bomb critical supply targets due to political restructions. Limited to interdiction of supply routes in S. Vietnam.	SEAD strategy responsive to technology. Initially restricted to Iron Hand strategies to get close enough to employ ARMs.
	No clear objective. No way of determining effectiveness of strategy.	
		Dec 72 destruction of IADS seen separate from SEAD; campaign used more than SEAD assets.
	Failed to recognize/respond to "air deniability" strategy of N. Vietnamese.	
Doctrine	AFM 1-1, dated 14 Aug 64, stressed strategic nuclear strike and deterrence concepts. Tactical offensive/ defensive considerations in conventional war treated lightly.	AFM 1-1, dated 28 Sep 71, stresses flexible response. Seeks to balance conventional/strategic options. Defense suppression/ electronic warfare noted as support roles.
Technology	Introduction of large-scale radar-directed threats. Pulsed radars, no jamming-protection circuits.	SA-2s increase frequency coverage, circuit design, and radar features to defeat missiles and jamming.

Table 1 (cont'd)

	1965	1973
	Introduction of AMG-45 Shrike (1964) ARM. Carried by F-100F to defeat SA-2. Short-range and limited flexibility.	Frequency coverage of radars expanded by addition of newer radars such as SA-3 Low Blow and higher frequency AAA radars.
	MiG-17/19 N. Vietnamese point defense fighters.	US responds by adding AGM-78 ARM and introducing EB-66 radar jammer and F-105G, increasing complexity of self-protection jammer pods.
	Rudimentary N. Vietnam communication net, very limited range, coverage, and redundancy.	
	Not critical to supply of Vietcong, hence shortfall not crucial to strategy of S. Vietnamese government.	MiG-21s with better IR and radar missiles and GCI control close out war.
		US responds by introducing special warning aircraft with specially developed detection equipment (still classified).
Political	Maintain viability of S. Vietnamese government by halting N. Vietnam-supplied Vietcong.	Maintain viability of S. Vietnam by destroying N. Vietnamese capacity to wage war.
	Honor treaty commitments under SEATO.	Turn war over to Vietnamese; gradually reduce active participation.
	Justified under basis of "domino theory" to protect national interest.	Domino theory largely discounted; political justification to protect national interest tenuous.
	Gulf of Tonkin Resolution gained strong support from Congress, public opinion.	
		Strong public opinion opposition; loss of support most directly correlated to US casualties.
Economics	War competed with LBJ "Great Society" plan. US in reasonably solid economic position.	US economic position precarious; off gold standard; dollar floats on currency market. Great Society in shambles.
	Congress continues to authorize funds without oversight into military expenditures.	
		Congress threatens to withdraw funding for war effort; exercises more and more oversight.
	War seen as a momentary drain on resources.	War costs set at several hundred billion dollars for nine-year commitment of forces.

Bekaa Valley

During the early 1980s, the Syrians, in support of their operations in Lebanon, constructed an elaborate IADS in the Bekaa Valley. Despite repeated diplomatic gestures from the Israelis stating they considered this an infringement of their sovereignty, the Syrians persisted in expanding their IADS structure.

The Bekaa Valley is located immediately to the north of Israel. The SAM sites were placed in such a way as to impede Israeli air operations projected into southern Lebanon. More seriously, they provided a corridor of access to Israel that the Syrians could use to attack.[18] With the bitter memory of the October 1973 Yom Kippur air war seared into their memories (the Israelis lost more than a quarter of their combat air force—150 aircraft—in the first three days of fighting to radar-directed AAA and SAMs),[19] the Israelis were loathe to permit a second occurrence of such devastating proportions. The Yom Kippur War had badly surprised the Israelis—especially the new SA-6 with its Straight Flush tracking radar and Gainful missile, which could avoid detection by Western electronic sensing devices or antiradiation missiles.[20] Oddly, the Israelis had lost only six aircraft in air-to-air engagements, a testament to their air-to-air acumen and an indication of the Arab league's poorly trained fighter crews. Even though the Israelis inflicted 456 aircraft losses on the enemy,[21] the Israelis resolved never to allow their air force to enter a war against such a lethal IADS net again.

On 6 June 1982, the Israeli Defense Force (IDF) commenced "Operation Peace for Galilee," as the Israelis dubbed the war in Lebanon. During the 1–10 June air portion of the campaign, the Israelis destroyed more than 20 SAM sites and shot down 90 Syrian MiG-21s and -23s in aerial combat.[22] The key to their success began, conceptually, where the war against the North Vietnamese IADS ended. The lesson was clear: the first order of business was to destroy the Syrian ability to defend its armies and territory from air attack. Several features of the Israeli air campaign in the Bekaa Valley stand out. Lt Gen Kelly H. Burke, USAF, Retired, pointed out that "Lebanon was the war of the future—a war in which electronic combat was a central and dominant theme."[23] The full range of electronic combat was used in support of this SEAD operation: ground-launched and air-launched drones, antiradiation missiles (air-launched and ground-launched), airborne jamming of communications and radars, and special detection and location equipment. More importantly, the SEAD operation, though primarily a function of electronic combat, was a concept that placed the entire military apparatus at its disposal in order to achieve the rapid, total destruction of the Bekaa Valley IADS.

The doctrinal shift from Vietnam—and the Israelis' own experience in the Yom Kippur War—was piercing. SEAD was not a support role performed by specific aircraft, nor was it a piecemeal application of assets to support a specific mission. The destruction of the Bekaa Valley IADS was the goal; a significant part of the total Israeli military apparatus was mobilized for the SEAD effort as an integral facet of the war.

A cursory analysis of the Syrian IADS in the Bekaa Valley at the time of the Israeli attack reveals a level of sophistication undreamed of 10 years before. Added to the Soviet-designed net used by the North Vietnamese and adapted for Syrian use were advanced, high-rate-of-fire AAA (ZSU-23-4) weapons capable of spewing in excess of 4,000 rounds a minute of accurately aimed 23-mm ammunition more than 9,000 feet with the use of their highly mobile gun dish radar firing control system.[24] Additionally, the Syrians possessed 13 squadrons of MiG-21 and -23 aircraft in their inventory which were tied to a redundant, highly cohesive command, control, and communications (C^3) net that used Soviet state-of-the-art radars for advance warning.[25] There were more than 20 SAM sites in the Bekaa Valley, consisting of SA-2s, SA-3s, SA-6s and even a sprinkling of SA-8s.[26]

Military analysis publications differ on the exact numbers and types of weapons the Israelis possessed at the outbreak of the Bekaa Valley campaign. A brief examination of Israeli acquisitions and improvements in the area of electronic combat illustrates their appreciation of its importance. As a ratio of defense expenditures, the purchase of electronic surveillance, detection, and antiradiation missiles increased at a much higher rate than other types of defense expenditures—even higher than the rates of the United States or the Soviet Union in the same time frame. So successful, in fact, were the Israelis in their improvements to US and Western European-built systems that Israeli armament industries sold their systems on the international market before and especially after the Bekaa Valley campaign. For a time, a variation of the Israeli harassment drone, the Scout, could have been a contender for the US antiradiation harassment drone program (Tacit Rainbow) as a complementary system to the AGM-88 high-speed antiradiation missile (HARM), based on advertising claims made for it in arms trade journals.[27]

The relationship of political objectives and economic factors to strategy and technology cannot be overlooked, for it is here that several strands in the variable matrix converge. Unlike US political-military mismatches of objectives and strategy in Vietnam, the *Israelis had clearly defined political objectives in their Lebanon operation, which were relevant and translatable to military strategy and amenable to the creation of tactics.* Defense Minister Ariel Sharon delineated three objectives. First, evict the Palestine Liberation Organization's (PLO) military and political organs from Lebanon. Second, engineer the election of Bashir Gemayel as the new Lebanese president (Gemayel being more receptive to Israeli policy than the Syrian surrogate). Third, bring about the withdrawal of the Syrian Army from Lebanon.[28] Curiously, in a rough parallel to US experience in Vietnam, the military appeared to have exceeded its mandate of 5 June 1982 which charged the IDF "with taking all the northern settlements out of the range of fire of the terrorists concentrated . . . in Lebanon."[29] Prime Minister Menachem Begin was specific in his instructions concerning Syrian Army troops, noting that they were not to be attacked unless they fired on Israeli forces. Depending, of course, on how one defines "terrorists," "range of fire," and "attacked unless fired on," one could argue that the language was sufficiently vague to allow the preemptive offensive action on the Israelis' part that commenced the following day. Since the air offensive was a thoroughly planned and practiced operation, one could easily conclude that the ambiguities were intentional.

A number of detailed analyses of the Bekaa Valley campaign exist. By piecing the various accounts together, one can gain a view of events as they probably unfolded, and gain a sense of the force structure utilized, the tactics involved, and the impact of various elements on the outcome of the SEAD campaign.

At some point prior to the opening of hostilities, the Israelis were faced with a series of strategic and tactical problems. As indicated, the largest strategic problem for their air force was the existence of an extensive and relatively sophisticated Syrian IADS in and around the Bekaa Valley. Syria's IADS capabilities allowed its early-warning system, located throughout the Bekaa Valley, on the Shouf Mountains and the border areas of Syria and Lebanon, to have a "deep look" into Israeli operations at medium altitudes. Total strategic surprise was therefore improbable. Nonetheless, if the Israelis were going to attack the IADS, they needed clear information about it. Probably for this purpose, the Israelis placed their electronic surveillance and jamming aircraft (four specially modified Boeing 707 aircraft plus E-2C Hawkeyes)[30] where they could listen and locate potential enemy radar sites without being placed at risk or appearing threatening to the Syrians.

A three-phase campaign ensued: the first phase was deception; the second, harassment with minor destruction (a continuation, in some respects, of the deception campaign); and the third, the actual destruction of the sites. Timing (time compression and sequencing), precision, and integration were critical. Presumably, through the use of tactical deception (habituating an enemy to see what you want the enemy to see and then doing something else), false targets (decoy drones and certain jamming techniques associated with equipment the Israelis possessed), and a series of false leads by attacking air forces, the Israelis might obtain effective tactical surprise. Massive, simultaneous, multiple-axis attacks might also have been employed because of the nature of the commitment to destroy the IADS target array. The combined effect of these tactics might have complicated an already overloaded and confused Syrian command structure's decision-making ability, making timely or correct decisions in allocating weapon resources against airborne targets increasingly improbable. These tactics would, essentially, have paralyzed Syrian IADS central control leaving firing units uncertain what they were shooting at—if they were shooting at all—and allowing attacking aircraft a considerable degree of immunity from the Syrian IADS weapons while at the same time delivering weapons on the sites.[31]

From the Israeli view, orchestrating such a large number of aircraft would have required precise timing and positioning of aircraft in order to perform their tasks in the correct sequence. The total numbers of aircraft involved in this effort and their relative contribution to the task is not known. It is clear, however, that the Israelis had decided that the destruction of the IADS was the primary objective of their air offensive and that they would commit whatever forces were necessary to accomplish the objective.[32]

The Israeli choice of weapons also seems to have shown a great ability to select the correct weapon for the correct target. The Bekaa Valley IADS' SA-2, SA-3, SA-6, and SA-8 radar-directed SAMs as well as a variety of infrared (IR) SAMs (SA-7, SA-9 vintage) were configured differently, and each had vulnerabilities which could be used against them.[33] These SAM systems were supported by radar-directed AAA (23 mm and 57 mm).[34] Completing the air defense net were the Syrian ground-controlled interceptors based in Syria only minutes away.[35] SA-2 and SA-3 systems, while transportable (can be dismantled, moved, and reassembled in 12 hours by a competent crew), are normally "fixed" sites and serve more strategic and point-defense purposes.

That being the case, the Israelis would not have had to employ sophisticated antiradiation missiles such as those used in the Wild Weasel role, where the site's location is not known. Nonetheless, the Israelis did use a large number of antiradiation missiles (presumably "to make sure") against location-known radars that continued to radiate.

Standoff weapons or free-falling bombs could have been used against these relatively static sites if any number of criteria were met. If the site was not radiating and AAA was not a factor, the site posed no threat to bomb droppers, thus alleviating the requirement for antiradiation missiles. In fact, these sites may not have been radiating for fear of being attacked by antiradiation missiles which home in on the electromagnetic emissions of the associated radars. Alternatively, the radar site may not have been radiating because its parent control center had not told it to emit (because the control center itself had been destroyed or jammed). The Soviet-designed Syrian systems, being quite old even in 1982, might also have had significant maintenance difficulties which kept them "off the air."

In the case of the SA-2s and SA-3s, another alternative tactic was employed to ensure a benign target environment. By selectively letting the Syrian IADS see exactly what they wanted them to see, the Israelis stimulated the SAM fire-control radars. Then they launched Samson decoy drones which looked to the Syrian radars very much like attacking aircraft.[36] The Samson drone, which the Israelis have marketed to the US Navy, is a lightweight, bomb-shaped drone with wings and a radar reflector fitted to mimic the radar cross section (RCS) of the aircraft carrying it. It glides at sufficient speed to decoy as an aircraft. Its purpose is to draw fire. The theory behind employing it is very simple: fill the sky with drones, force the enemy to expend all SAMs, and while the adversary is reloading, attack in force from an unexpected direction.

The tactical problem the SA-6, SA-8 and ZSU-23-4 systems pose is that they are mobile. They can be moved at a moment's notice so that targetting information that is only hours old may be useless. Curiously, the Syrians obliged the Israelis by positioning their SAMs—including the SA-6s—in fixed-site configurations. From a weaponeering and tactics viewpoint, the Israelis approached this problem in several ways. First, by utilizing real-time intelligence gathering, the Israelis were able to locate mobile systems by using a remotely piloted vehicle (RPV) mounted with a camera and a transmitter (the Mastiff) that allowed the battlefield commander instantaneous view of areas of interest.[37] By acquiring such data, they could use the same tactics that were successful against the older SA-2 and SA-3 systems. Second, the Israelis used the electromagnetic signature of the emitter itself to locate the emitter with equipment designed to detect, identify, and locate the specific Syrian SA-6, SA-8, or ZSU-23-4 radars, no matter where they were on the battlefield (providing they could get the radar operators to emit). Specially modified antiradiation missiles designed to track these specific radars were fired from both airborne platforms and probably from prepared ground-launch sites.[38] These systems were especially effective during the first stages of the campaign when all radar sites were radiating.

Radar jamming was used in a variety of ways to support the operation. First it was used to blind selected corridors of attack. Then it was used to deceive the Syrian early-warning and acquisition radars. By jamming a specific area, the Israelis were able to obligate the Syrian IADS into committing forces against a phantom attack force. The result of alternating between real and phantom corridors was catastrophic to the Syrian SAM battery commanders and sector controllers.[39] Third, a combination of selective communications jamming and radar jamming enticed Syrian fighters tied to GCI controllers into an ambush. The Israelis, once again, apparently allowed the Syrian IADS to see what they wanted them to by allowing the Syrians to direct fighters to an area where they would be vulnerable to Israeli attack (a "kill zone"), and then initiating jamming to sever the communication links with their Syrian GCI controllers. Because the Israelis knew how dependent Syrian fighter pilots were on GCI, severing the Syrian link in this way resulted in the Israelis' shooting down 90 Syrian aircraft without a single loss.[40] The IDF also aided in the SEAD campaign by providing selected helicopter assault and artillery fire against some of the closer SAM sites in the southern Lebanon area and by launching both ARM and RPV assets to support the effort.[41]

The combined effect of these tactics in support of the SEAD campaign strategy was devastating. *After the first hour of combat, the overwhelming majority of the Bekaa Valley IADS lay in ruins.* Repeated forays into the area prevented the Syrians from ever reacquiring a defensive initiative. With their ability to defend against air attack destroyed, the Syrians' will to press the fight evaporated. More important was the latent threat that the Israelis could repeat the performance on demand against the Damascus homeland's IADS and lay Syria proper open to unopposed aerial attack. The Syrians understood the latent threat and responded by withdrawing the remainder of their forces.

What is clear from the Israeli experience is that they used the assets they had on hand to accomplish their purposes. What they did not have, they improvised. The strategy was driven by the ultimate objective of the campaign—the destruction of the Bekaa Valley IADS—not by the technology or the doctrine. US strategy and tactics are often seduced by technological gimmickry and often substitute the technological quick fix for a lack of clearly defined military objectives. Also, political objectives as filtered through existing doctrinal and strategic concepts are not generally translatable into usable military tactics.

Two notes of caution are in order when using the Bekaa Valley as a model of the future—which, for the most part, it became. First, the Bekaa Valley campaign—as successful as it was—was conducted only after exhaustive practice, only after precise intelligence requirements were met, and only when the needed material was on hand to accomplish the task. The initiative belonged to the Israelis from beginning to end. Second, the environment of the desert was uniquely suited to the weapons used. The same systems and tactics would not have worked as well in a damp, forested, hazy environment, or in one where the enemy was able and/or willing to mount a substantial counterstrike. Both these caveats obtain in much of the world where future American forces may be employed.

The Bekaa Valley matrix reveals notable changes in emphasis and direction when compared to the Vietnam matrix. Although the Bekaa Valley campaign was not an American experience, its impact on tactics and electronic combat in US operations in Libya and Saudi Arabia is undeniable. SEAD had become the initial, primary task to be performed—even before the traditional "air superiority" mission was executed. In some ways, in fact, air superiority was subsumed under SEAD as a function of defeating the interceptor portion of the enemy IADS. This definitional growth of SEAD may be presumptuous, but it serves to illustrate the growth of SEAD from a defensive, protective, reactive tactic to a campaign objective in which all other missions and roles contributed until the task was accomplished.

Table 2 shows the results of evolution across the 10 SEAD variables when compared to Vietnam. One can begin to sense the growing meld of technology and tactics in the "chess game" for air superiority or air deniability and the movement toward more encompassing concepts of SEAD.

Table 2

Bekaa Valley Matrix
Operation Sea of Galilee: June 1982

Threat	Bekaa Valley IADS consisted of 19+ SAM batteries comprising SA-2, -3, -6, and, after 10 Jun 82, some SA-8 batteries. These were supported by ZSU-23-4 gun dish batteries as well as optical AAA and older radar AAA.
	A 450-combat-aircraft Syrian Air Force consisting of late-generation Soviet fighters netted to the latest generation GCI air defense network.
	Long-range, modern Soviet-designed radars capable of detecting operations throughout most of Israel at medium/high altitudes.
Definition	SEAD is functionally defined according to military objectives and needs. Objective of operation was access to terrain via vehicles and air. Therefore SEAD defined as campaign to facilitate objective. Defensive/offensive aspects defined accordingly.
Tactics	Coordinated plan based on deception and total force concept. Army, Air Force, Navy assets directed by single individual. Preplanned aspects, plus dynamic assets, integrated.
	Tactical deception using drones, feints, and jamming utilized to "bait" IADS.
	Offense-oriented. Stressed objectives of campaign tied to strategy.
	Jammer assets (communication and radar) coordinated in creating kill zone for Syrian interceptors.
	ARMs mixed with drones and fighters just in front of main attack force. Complicated SAM operator's decision-making process.
	Simultaneous use of drones, ARMs, fighters, and selective jamming of principal early-warning radars paralyzed and/or deceived IADS controllers.
	Thoroughly rehearsed and practiced.
	Time compression and suddenness of attack critical.
Organization	Centrally controlled, centrally executed. Command and control links directed from single location by single commander using unified force structure plan.
	Clear, protected command and control links.
	Endemic to total force. No specialized SEAD organizations.
Force Structure	SEAD assets distributed throughout Army, Navy, Air Force. Assets dedicated to Bekaa Valley campaign consisted of air/ground launched AGM-45/78s, air/ground launched drones (decoy, etc.), standoff jammers, field communication jammers, real-time intelligence, remotely piloted vehicles.
	SEAD campaign configured forces for "one-time" execution of specific plan. No permanent SEAD structure resulted.

Table 2 cont'd

Strategy	Offensive ground/air blitzkrieg with specific military objectives. Destroy Syrian ability to defend themselves from air attack, force withdrawal of Syrian Lebanese forces.
	Designed to achieve quick victory because of uncertain political support.
	SEAD campaign designed to protect entire force as opposed to defense of attacking aircraft. Relied on self-protection of aircraft and maneuver tactics for individual aircraft defense.
Doctrine	To gain air supremacy over battlefield using whatever means necessary. SEAD subsumed under overall air supremacy concept.
Technology	Major improvements in electronic receiver technology and miniaturization of computers allowed greater flexibility in locating/targeting radar emitters.
	Extensive use of drones and RPVs extended visualization of battlefield.
	F-15/F-16 proved more effective against MiG-21/23 in radar and IR missile engagements.
	Technological advantage shifted to offensive SEAD weapons if used in concert with adequate strategy and executed with proper tactics.
Political	First war fought as a "power projection" war outside of Israel without Israel having been attacked.
	Military objective (withdrawal of Syrian forces from Lebanon) clear, but other political objectives not clearly influenced by military-only means.
	Clear military victory via SEAD campaign but uncertain political outcome.
	Military activity exceeded stated political mandate; support for military strike ambiguous.
	US/international response not universally supportive; much opposition. Isolated Israeli government, Defense Minister Sharon ousted.
Economic	Resource-restricted budget. Dependent on US aid and favorable loan treatment to obtain military hardware.
	Many workarounds in hardware due to fiscal constraints.
	Because of political problems, future major military requirements uncertain.

Libya

State-sponsored terrorism spawned by a number of middle eastern nations was responsible for the deaths of several hundred people worldwide during the mid-1980s. One of these perpetrator nations was Libya. Because of Libya's flamboyant leader, Muammar Qaddafi, the state policy supporting terrorism remained boisterous and unrelenting. Libya provided training, refuge, and materiel for these terrorist operations. United

States citizens were a primary target of many of these activities, and in March 1986 the Reagan administration issued what amounted to an ultimatum to the Libyans. In early April 1986, a terrorist bomb exploded in a German nightclub that catered to American servicemen. It killed several people, including an American serviceman. It cannot be ascertained whether this incident occurred because of a blatant disregard for the American warning, a falsely held belief that the Libyan air defense network—buoyed by years of extravagant spending on the latest Soviet IAD systems—made them invulnerable to attack by the Americans, or a belief that the United States was bluffing. What is certain is that the US did respond. During the early-morning darkness of 15 April 1986, an armada of US Air Force aircraft that were launched from the United Kingdom flew around the Iberian peninsula through the straits of Gibraltar and attacked the city of Tripoli while elements of the US Mediterranean Fleet simultaneously attacked the city and surrounding area of Benghazi to the east. The attack lasted 11 minutes. The targets were precisely defined prior to the raid and explicit instructions were given to attacking aircrews on the rules of engagement to be followed. The raid had been practiced in bits and pieces prior to the attack, following the Israeli experience of the Bekaa Valley. The equipment required to accomplish the mission, the routes to be flown, the composition of the forces, and the tactics to be employed had all been predetermined and exhaustively planned.

Changes in technology—especially electronic technology—played a major role in the Libya raid. The changes were large enough to create a qualitative difference between how SEAD was accomplished during the Libya raid and during the Israeli experience in the Bekaa Valley. These differences revealed themselves in several ways. Intelligence gathering and advance strategy/mission planning were greatly enhanced by computer application. The use of advanced high-speed antiradiation missiles (the AGM-88 HARM)—especially during the Benghazi raid—made direct attack of the sites largely unnecessary. The numbers and sophistication of radar-jamming aircraft available altered the way tactics were applied.[42]

Mission planners made extensive use of the latest technology. Computer simulations of the Libyan IADS, fed by information that was—many times—only minutes old via worldwide satellite communications, were developed to show the best routes to fly to evade the detection and lethal envelopes of Libyan air defenses. Unlike computer models of the immediate post-Vietnam era, these models represented the "most likely case" scenario, as opposed to the "worst-case" scenario popularized throughout the 1970s and early 1980s. Military intelligence experts had now reached a point where their confidence in the data they displayed did not require the "what-if-we're-wrong-and-the-enemy-is-stronger" reflex action of the worst-case scenario. Computer models also aided in determining how many and what type of weapons would best accomplish the task. American aircrews entered the fight with an excellent idea of what they faced and with the confidence that they could overcome the obstacles placed in their path by a formidable Libyan IADS.[43]

US Navy aircraft employed the HARM since the Germany-based F-4G aircraft were not utilized in the raid.[44] HARMs could be launched well outside the lethal envelope of the intended SAM target. So long as the site continued to radiate, the missile would guide on the radar. This allowed mission planners to place HARM missiles on SAM sites immediately prior to friendly aircraft penetrating the lethal envelope of the SAM battery without endangering the attacker or the HARM shooter.

The employment of the EF-111 and EA-6 aircraft, however, was the key difference between Israeli and American SEAD tactics. Radar jamming of the entire IADS, as opposed to destructive means of suppression or piecemeal jamming, was to play a larger role than it had for the Israelis in 1982.

American SEAD tactics of the period were still threat-based, though the threat was seen in more systemic terms than the one-on-one proposition that characterized SEAD operations in the Vietnam War. The basic fabric of attack was balanced between achieving the objective by the most direct means possible and the requirement to protect friendly aircraft from destruction by the IADS.

Largely because of the unpopularity of the Vietnam War, a growing unstated objective of American combat became driven by political concerns. This objective has emerged as a corollary to committing American forces to armed action: the requirement to limit American casualties. As a result of this largely political undercurrent, SEAD's importance to overall military objectives seems accentuated by its emphasis on reducing attrition. Further proof of the effects of this phenomenon's evolution is found in Desert Storm. The corollary to reduce American casualties in the pursuit of limited objectives played a significant role during Desert Storm in decisions to employ SEAD as a primary campaign objective.

A brief description of the Libyan IADS at the time of the raid reveals a concentration of radars, AAA, and SAM density the equal of any in the world. The quality of their air force was suspect, though they possessed frontline Soviet aircraft.

What is not widely appreciated among historians and strategists is the proliferation of Western technologies to potential adversaries. The Libyan IADS consisted not only of Soviet-designed equipment but French, German, and English systems as well.[45] Because these systems were considered "friendly," there was little knowledge on how to counter some of them. Additionally, the Libyans adapted many special-purpose radars. For instance, British, German, and Russian manufactured naval surface-search radars were drafted to serve as coastal early-warning radars. These radars performed quite well and posed some unique jamming problems to the EF-111 and EA-6 aircraft.

The lesson of the Bekaa Valley for the Libyans—more precisely for their Soviet advisors—was to create a credible technical IADS structure to counter known US strengths. The task involved four major challenges. First, the Libyans were compelled to build a system with more density of coverage where the destruction of one radar would not leave a gap. Second, they were obligated by the existence of American jamming aircraft to field radars with greater diversity of frequencies where the jamming of one operating frequency would not disrupt total early-warning, SAM, and AAA operations. Third, the Libyans ascertained that by employing a greater variety of systems utilizing differing electromagnetic waveforms they might be able to complicate and confuse attacking aircraft's detection and electronic self-protection equipment. Fourth, through creating redundant C^3 links by building hardened landlines as well as multiple radio links (since the Libyans adhered to Soviet close-control doctrine, this was especially important to them), they might complicate American plans to decapitate the decision-making structure. As can be clearly discerned, the technological chess game continued to accelerate in the SEAD dimension.

A brief analysis of the Libyan IADS shows that they placed their defensive sites mostly toward the sea, the presumption being the aircraft would not attack from the landward side. They placed their most capable systems along anticipated corridors of attack, and even employed extremely long-range SAMs. AAA systems were sprinkled throughout Tripoli and Benghazi in overlapping coverage. Early-warning radars, GCI radars, fire-control acquisition radars, and target-tracking radars were netted together through an intricate web of redundant systems and links. Several fighter bases were within minutes of the target areas and, given appropriate warning and direction, these aircraft could be launched at a moment's notice.

Unlike the Bekaa Valley campaign, the Libyan SAM sites and interceptor air fields were not the primary targets for the raids on Tripoli and Benghazi. The raids were designed as a measure to punish Libya for sponsoring terrorism and to deliver a strategic blow at terrorist training conducted in the area. As a result, radar jamming could be used as a primary means of degrading the overall performance of the Libyan IADS as American fighter-bombers needed only to have the IADS neutralized for the period of the attack.

In some cases, because jamming aircraft did not possess the correct frequency coverage for the specific intended radar or because destructive means of removing the threat were judged more effective, the attackers employed a combination of antiradiation missiles and direct bombing to suppress specific sites (shades of Iron Hand and hunter-killer tactics from Vietnam).[46] This was especially true for the Navy attack on Benghazi.[47]

The raid on Libya marked the incipient stages of a deliberate effort to combine Navy and Air Force SEAD assets in the prosecution of a battle. The bounds of SEAD expanded again—this time to assume the complexion of an integrated force operation. The Libya raid also provided a comparison of Air Force and Navy applications of SEAD and insight into the tactical, strategic, and doctrinal vantage points from which differences in SEAD employment since Vietnam SEAD sprang.

Navy SEAD doctrine was—and is—largely unwritten. The roles naval SEAD plays are tied directly to missions of the Navy—these being sea control, strategic sealift, and power projection ashore. The corollary to these is an exceptionally strong emphasis on fleet defense in comparison to either Air Force or Army doctrine. The Navy's attack on Benghazi and their support of Air Force operations on Tripoli illustrate the tactics involved, but no strategy per se can be gleaned from their operations. Clearly, the Navy was concerned with power projection and fleet defense.

The primary differences between Navy and Air Force applications of SEAD stemmed from two sources. First, the Air Force, in the Libyan operation, was resource-limited and could not deliver fighter cover or Wild Weasel support because of fuel constraints, security concerns, and political considerations. Second, Navy SEAD concepts dictated that they employ their jamming and antiradiation missile assets differently, even though the equipment was similar to Air Force assets. A comparison of US Navy and Air Force jamming aircraft used in the Libya operation—the EA-6 and the EF-111—is illuminating.

The EA-6 uses essentially the same jamming subsystem employed by the EF-111; both are variants of the ALQ-99 jammer. The EA-6 is an aircraft more than 10 years older than the EF-111 and is itself a variant of the Navy's primary heavy attack aircraft, the A-6. Even though the aircraft share similar systems, only in the power-projection role do they have similar missions and tactics, and even here, the tactics of the EA-6 reflect its blue-water heritage. The jammer the EA-6 employs has much greater computer capacity, much more sophisticated jamming techniques, and response to a wider variety of radar types to include early-warning/GCI, acquisition, and specific fire-control/target-tracking radars. The EA-6 has the capability to fire HARMs, though in a way much different than an F-4G. It is designed to protect the fleet, fight the war at sea, and project power.

The EA-6 system is versatile, but it sacrifices specific capabilities to achieve this versatility. While the Navy would doubtless prefer to use F-4G-type tactics in employing the $200,000-plus HARM, it has neither the space aboard its carriers to support the addition of such a specialized aircraft nor the fiscal resources to alter existing aircraft to perform the Weasel role. A naval air wing's aircraft must be multiroled for efficiency.

Nonetheless, the ALQ-99 and the formidable HARM are mated to an EA-6 aircraft with subsonic speed, 1950s-vintage aerodynamics, and limited range. This restricts the flexibility of the aircraft in that it cannot escort higher speed fighters into a target area and it is limited by the range to which it can penetrate into enemy airspace without aerial refueling. In its power-projection role in Libya, the EA-6 performed limited escort of Navy fighters to the edge of the Benghazi target area where it assumed an orbit outside the lethal range of enemy systems. The EA-6 also performed with the EF-111 in strategic jamming of the Libyan primary early-warning/GCI radar grid from a standoff orbit providing support for the Air Force attack on Tripoli.[48]

The EF-111 is also a variant of an existing aircraft, the F-111A. It is a high-speed (Mach 1+ at sea level, Mach 2+ at medium and high altitudes), all-weather, day or night fighter that retains the F-111's low-altitude, terrain-following features. It has greater absolute jamming power than the EA-6 and greater unrefueled range. Whereas the EA-6 has a crew of four, the EF-111 has a crew of two. The ALQ-99 subsystem employed by the EF-111 in the Libya raid was not as sophisticated as that employed by the EA-6, but for the purposes of the raid, it was more than sufficient for most of the targeted radars.[49]

For the Libya raid the US Air Force appears to have opted for jamming as the primary element in their SEAD operation. We can deduce this from three features of the attack. First, the primary attack aircraft needed only to evade the SAM sites to accomplish their primary bombing mission. Second, the low-altitude, high-speed, night attack would limit the radar horizon. At 200 feet, an attacking aircraft would be first detectable at about 25 miles (assuming the radar is at ground level) because the curvature of the earth blocks the radar's line of sight. At attack speeds, given no advance warning, an enemy radar would have less than three minutes to locate, identify, track, and allocate a weapon before the attacker would be overhead the site, even if no jamming were present. Third, the primary radars the Libyans employed were composed largely of older generation radars for which the EF-111 and EA-6 jamming systems were optimized.

Jamming the older generation of radars that comprised the bulk of the IADS would deprive the Libyans of information with which to cue their fire-directors and air defense sites. More importantly, because of Soviet-style command-and-control doctrine, no information equated to no decision. Since the Libyans were not trained in autonomous operations (operating without instructions from above), a "no-decision" order would result in the general collapse of their air defense net.

The nonlethal jamming SEAD option was extremely effective. Interviews with several El Dorado Canyon participants revealed that, to their knowledge, not one F-111 was acquired by enemy target-trackers, nor were any enemy interceptors observed.[50]

The Navy's use of HARMs and EA-6s on the Benghazi attack proved equally effective. As noted before, the US Navy does not have a pure Wild Weasel-type aircraft, though many of their aircraft have the capability to launch the HARM (the A-6, A-7, EA-6, and F/A-18). The Wild Weasel, with its specialized and very sensitive APR-38 radar detection, identification, and location gear, was able to distinguish if a specific radar was operating, where it was, and what it was, and to relay this information directly to the guidance section of the HARM it was carrying. The HARM was (and is) a resource that needed to be used as efficiently as possible because of its relatively limited numbers and great cost. Because the Navy had the missile but not an APR-38-equipped aircraft from which to launch it, their tactics required the aircrews to fire multiple missiles at a general target area in a preemptive mode. These aircraft have limited to no capability to detect whether their intended target is radiating. The Navy concept is straightforward: if a radar comes on the air, one of the missiles will guide toward it. If the sites under attack practice emission control and shut down their radar, the missiles have still accomplished their purpose of suppressing the radar by keeping it off the air. Discussions with Navy HARM shooters reveal that the cost of the missile (and their—from the US Air Force's view point—inefficient use of them) was not their concern; they had the missile and this was the most effective way they had of using it. The raid was a one-time shot, and the use of many HARMs to take down a handful of SAM sites was perfectly justifiable in their eyes. Exhausting the inventory was not a serious concern. Besides, as one Navy pilot indicated, the supposed inefficient use of the HARM depends on how one defines efficiency in combat.[51] The Libya raid experience presented two major evolutionary features for SEAD when compared to the previous benchmarks (see table 3). One was the continued, deliberate, and integrated use of Navy and Air Force assets to accomplish a single purpose as opposed to dividing up the turf and fighting separate wars. While the Israelis had melded their services together brilliantly in achieving their successful Bekaa Valley SEAD campaign, they did not do it as a function of force projection from great distances. Also, SEAD was an integral part of the attack plan, not a separate phase. The Israeli SEAD attack was a separate phase that, once accomplished, enabled unobstructed access to the land and air. The American use of SEAD was very specifically tailored to the tactics and strategy that supported the objectives of the raid.

Table 3

Libya Matrix
Operation El Dorado Canyon: 15 April 1986

Threat	Qualitatively/quantitatively different from Vietnam/Bekaa Valley. Use of Soviet and Western-designed radars and air defense weapons systems unique. Increased importance of intelligence role for SEAD to acquire data.
	Density of threat equivalent to East Europe, but with many systems more advanced and technologically capable than Soviet-only systems.
	Tactics/strategy heavily influenced by nature of threat.
Definition	Functionally directed at radar components of IADS as a function of protecting friendly attackers.

Table 3 cont'd

Tactics	Significant changes since Vietnam. Use of specialized assets to accomplish SEAD. "Electronic combat triad" established as core of SEAD: F-4G, EF-111, and EC-130H Compass Call communication jammer.
	Political/economic/logistic constraints limit triad to EF-111. Navy EA-6 and other HARM shooters complement EF-111 jammers and perform quasi-Wild Weasel role.
	Timing key to success. Jamming commences just prior to strike, immediately by launch of HARMs. Maximum confusion, indecision for Libyan controllers.
	Navy EA-6/HARM shooters folded into USAF SEAD tactics scheme. USAF Wild Weasel planners/analysts critical of "wasteful use" of HARMs.
	EA-6/EF-111 proved very compatible/complementary during raid.
	Complexity of threat forces—like Bekaa Valley—rehearsal of key features of SEAD attack plan.
Organization	As function of electronic combat, 66th Electronic Combat Wing, 65th Air Division, 17th Air Force, USAFE, formed consisting of all USAF European-based electronic combat assets (EF-111s at RAF Upper Heyford, F-4Gs and Spangdahlem AB, Germany, and EC-130Hs at Sembach AB, Germany). Separate squadrons of EF-111s, F-4Gs, EC-130Hs under 12th Air Force, Tactical Air Command. Pacific Air Forces maintained a squadron of F-4Gs in the Philippines.
Force Structure	Epitome of SEAD as specialized electronic combat function. Force structure mirrored tactical division of labor at time of Libyan raid.
	Navy electronic combat assets supporting AF assets folded into USAF force structure/strategy scheme.
Strategy	Intent was to use SEAD to prevent loss of US air assets to SAM/AAA. Resulted in neutralizing SAM/AAA as well as the GCI threat. Implication toward use of SEAD as element of offensive counter air.
Doctrine	Subtle shifts in doctrine. Tactical use of SEAD had significant impact on strategic campaign. Integrated use of Navy/Air Force electronic combat assets affect both airland battle concepts and power-projection ashore.
Technology	Threat driven primarily by technological advances. Increased power, frequency diversity, electromagnetic waveforms of IADS radars complicate US tactical planning. Entered raid with some "best guess" measures and workarounds.
	For first time, IADS assessed as capable of night operations. For variety of reasons, Libyan IADS proves completely ineffective.
Political	Strong backing of American people, reluctant backing of Europe, strong negative response from third world. American Barbary Coast action of 1803, and the Pedicaris affair of 1905 provided sufficient precedent for American action against international terrorism.
	Achieved desired effect. Raid nature of attack matched punitive objective of strategy.
Economic	Multifaceted influences and effects. Fear of oil embargo froze some European political support of operation. Cost of equipment and military operation prohibitive for extended military activity. Cost of specialized assets and electronic equipment required to neutralize threat increasing faster than perceived payback. DOD began to explore more cost-effective SEAD options.

The other evolutionary feature was the effect from the balance of lethal and nonlethal means of SEAD. Jamming, as a primary element of SEAD, began to assume a more prominent role. The Bekaa Valley attack used jamming as a means of obscuring attack intentions in order to destroy the sites. While it is difficult to compare Libya and Bekaa Valley on this account because the target objectives were different, it seems probable that the Israelis would have opted for destruction of the sites under any conditions as a function of doctrine. American SEAD tactics evolved with the recognition that a disabled enemy IADS, whether by jamming or destruction, is the object of SEAD, and if jamming can accomplish the task more effectively, efficiently, and safely than destructive means, then the jamming option should be chosen.

The matrix input for the Libyan operation allows us a broader comparison than before. Clear trends, areas of convergence, and interior relationships among the variables emerge with greater clarity.

Persian Gulf

Many of the studies concerning Operation Desert Storm are not yet complete. In fact, the Air Force's Studies and Analysis Branch at the Pentagon has amassed 1,200 separate studies and after-action reports (which they call "data points") comprising several hundred thousand pages, and this is only the beginning.[52] Separate studies are being conducted at Tactical Air Command (now Air Combat Command), Air University (now a part of the Air Education and Training Command), the Air Force Electronic Warfare Center, and the Joint Electronic Warfare Center; and a separate Desert Storm Study Office has been chartered by the secretary of defense. For such a public war, however, surprisingly little data is available on SEAD's contribution to the effort. A great deal has been written about the Persian Gulf War, but much of it is a rehash of bits and pieces of the same data—not new data.

This problem is especially vexing for an analysis of SEAD's evolution. The data available is incomplete. That we won the war in dramatic fashion does not, in itself, vindicate the tactics or all of the strategies involved, nor does it establish the doctrine that brought this victory about as the final word. However, Desert Storm does represent a unique opportunity to explore a process as it takes place in much the same way as a biologist examines a living organism and compares it to the fossil record.

During the night of 2 August 1990, Iraqi armed forces invaded the country of Kuwait. After a brief struggle, that country's small forces were crushed, and Kuwait was occupied. Saddam Hussein declared that Kuwait had been reclaimed and that it was always part of Iraq. This subterfuge disguised the fact that Iraq owed Kuwait more than $20 billion it borrowed during the devastating Iran-Iraq conflict that had ended indeterminately. There were also disputes concerning a major oil field that overlapped Kuwait and Iraq.

From a military point of view, Iraqi forces were poised on the doorstep of Saudi Arabia. Hussein's acquisitive taste for oil-bearing real estate appeared insatiable and belied his stated intentions with regard to Saudi Arabia. On 3 August 1990, elements of the USAF 1st Tactical Fighter Wing (Langley AFB, Virginia) were deployed to Saudi Arabia at the request of the Saudi government. Curiously, there is little these aircraft could have done in a substantive way other than to show American resolve. The 48 F-15Cs deployed were air superiority fighters with no air-to-ground weapons delivery capability. But these were followed within two weeks by 26 F-15Es (air-to-ground, two-seater variants of the F-15) and 48 F-16Cs, which can perform air-to-ground and limited air-to-air roles. Shortly thereafter, the Saudi government invited a huge international force into their country to protect the precious Saudi oil resource and to provide a base of operations from which to evict Saddam Hussein's now firmly entrenched military in Kuwait. Operation Desert Shield, the deployment operation that preceded Desert Storm, consumed almost six months of intense logistics activity in order to prepare for active military action and was unopposed. Apparently, Iraq preferred to adopt a purely defensive posture. In all, more than

2,000 combat aircraft were stationed in or around the Arabian Peninsula to support what would become Desert Storm.[53]

As a coalition leader, the United States decided around November to double the size of American forces in response to a perceived increase in the buildup of Iraqi forces in Iraqi territory adjacent to Kuwait and northern Saudi Arabia. By mid-January 1991, more than 500,000 Americans were committed to the Desert Shield effort along with more than 200,000 military members from the other coalition forces. With the sanction of the United Nations and the consent of the US Congress, Operation Desert Storm commenced in the early-morning hours of 16 January 1991.

The catalogue of forces was impressive. More than 46 percent of Tactical Air Command's assets were deployed. Almost 100 percent of the Marine Corps' and a very sizable portion of the Navy's air power were deployed. Significant levels of munitions, aircraft, and personnel were borrowed from the United States Air Force Europe and the Pacific Air Forces. The overwhelming bulk of Military Airlift Command's C-141s and C-5s were committed to the Persian Gulf theater of operations as well as the bulk of Strategic Air Command's tanker forces.

Sheer numbers and overwhelming firepower, however, do not always ensure victory. The quality of the equipment, the appropriateness of that equipment to the task, a thorough knowledge of the enemy's strengths and vulnerabilities, and a plan which matches resources to specific, clear objectives in accomplishing overall strategy are required to triumph on the battlefield. Above all, the soldiers who were deployed had to be well trained, committed, and well led.

That the Iraqi capacity to wage war was crippled so swiftly and completely belies the great complexity and severe challenge they actually posed for coalition air forces. The Iraqi integrated air defense system was in many ways the finest in the world. Netted together with an extremely sophisticated system of landlines and microwave systems as well as standard relay radios, the Iraqi IADS was fed by a system consisting of literally thousands of radars and observation posts.[54] The radars were some of the most advanced in the world, including the latest family of French surveillance radars as well as radars of modern Soviet design. They even had tried to adapt a French land-based radar for airborne surveillance purposes.[55] The net included radars and communications equipment of German and English design. The C^3 structure was assessed to be the most hardened in the world after the former Soviet Union and one of the most capable. At the heart of the C^3 system, then, was an Iraqi system that probably combined Soviet with Western European systems and developed and adapted it with doctrine forged in waging the eight-year war with Iran. Like the Libyans', the Iraqi system was an extremely capable hybrid of Iraqi, Soviet, and Western design. An accurate intelligence assessment of intentions, capabilities, and concept of employment was both critical and extremely difficult.

The Iraqis employed a full range of air defense weapons. Prior to the start of Desert Storm, they possessed the sixth largest air force (750 combat aircraft) in the world with some of the most advanced aircraft—including the latest French F-1 variants, the latest Soviet fighter—the MiG-29 Fulcrum, the MiG-25 Foxbat, and a plethora of other fighter and ground attack aircraft. They possessed between nine and 10 thousand AAA pieces and as many as 17,000 SAMs consisting of SA-2/3/6/8/9, Roland, and even the American I-Hawk—presumably stolen from Kuwait.[56]

The more important question, then, is, given the fact the Iraqis possessed formidable resources and a very competent air defense doctrine, why did they not acquit themselves better? Like the events of the war, the answer is complex.

Most of the answer is found in how the coalition's forces acquired and used intelligence compared with their Iraqi counterparts. Immediately after the Iraqis attacked Kuwait, US intelligence agencies expended enormous effort to learn as much about the Iraqi air defense system as possible. The Joint Electronic Warfare Center published detailed reports concerning the disposition of the Iraqi IADS and updated it as events required. Not only did intelligence agencies gain a deep appreciation for the IADS, but they transmitted this knowledge to a decision-making structure that understood the implications of tackling such a sophisticated IADS without sufficient protection. This situation was markedly different than Vietnam. In comparison, the Bekaa Valley intelligence-gathering effort and the Libyan operation were more limited in scope due to the limited nature of the objectives. The primary difference was the all-encompassing nature of the intelligence-gathering effort. The most important difference among the four historical benchmarks in this respect, however, had to do with the outputs and analyses of the intelligence data. In the previous campaigns, the data was used to target specific systems or avoid exposure to attacking aircraft. In the Iraqi assessment, intelligence was used to identify critical weaknesses which could be exploited for the express purpose of causing maximum damage to the Iraqi structure with the smallest expenditure of resources. In electronic combat, this process is known as "critical node analysis."

Critical node analysis was made possible by several factors. First, the technology was available in the form of electronic surveillance of enemy emitters that could relay data to decision makers in nearly real time. Second, through the application of improved computer technology and software programs, decision makers could visualize the way the enemy worked and would likely respond to a variety of proposed coalition tactics. Third, with the development of specialized aircraft and tactics, the results of analysis could be transferred readily to reality rather than be condemned as wishful thinking (as many SEAD studies of the 1970s and early 1980s were). In short, even though the Iraqis possessed a formidable net, they were unable to translate technological capability to purposeful action because they did not fully understand the system's capabilities and because US SEAD strategy was aimed at ensuring they could not use their technical capabilities even if they did.

Throughout the evolution of SEAD runs a deeper current that is much more subtle. It does not become apparent until Operation Desert Storm. Beginning in the late 1970s, the Air Force began a concerted effort to integrate and apply electronic combat to day-to-day training. Much of this training was traditional and concentrated on the use of self-protection equipment used in the event an aircrew should find itself under attack by a hostile target-tracker. More importantly, the training was directed at developing a more encompassing concept of waging war against the enemy using integrated electronic combat as the focus of a coordinated air-land-sea attack against the totality of the enemy's force structure.

By 1981, with the advent of the Air Force's "EC triad" (the F-4G ARM shooters, the EF-111 EW/GCI radar jammers, and the EC-130H communication jammers), the US Air Force was in an excellent position to orchestrate such an attack. Unfortunately, much of the tactical air force was unfamiliar with what electronic combat could provide for them and they habitually failed to include these aspects in their overall tactical planning. Beginning in 1981, the USAF Tactical Air Warfare Center, at Tactical Air Command's direction, established Exercise Green Flag, which stressed the importance of electronic combat to a variety of aircrews in a realistic warfare environment over the air gunnery ranges of Nellis Air Force Base in the Nevada desert. In the Pacific Air Forces (PACAF), Cope Thunder exercises provided the same sort of training in the Philippine Islands, and, in the late 1980s, the creation of a large electronic combat range in France and Germany allowed USAF Europe (USAFE) aircrews the same opportunity. Through the course of 10 years of integrated electronic combat exercises, thousands of aircrews experienced coordinating with electronic combat assets, especially SEAD assets, in the prosecution of their tactics. In fact, many of the aircrew who served in Desert Storm indicated that without the experience gained in exercises like Green Flag and its counterparts in Europe and PACAF, their individual flight tactics would have been much less effective.[57]

It was the intrusive influence of electronic combat—and its SEAD stepchild—that, perhaps, enabled a smooth transition from the piecemeal, one-on-one application of SEAD in Vietnam to the total-force application of SEAD in Desert Storm. It is this integrated, total-force application that will serve as the basis for "true" joint SEAD operations of the future.

The overall strategy of Desert Storm also was influenced by the use of the total-force concept to employ SEAD. The Desert Storm air campaign plan, as originally intended, was to be accomplished in four phases:

1. Gain air superiority.
2. Suppress enemy air defenses in Kuwait.

3. Keep the pressure on Phase I and II targets while shifting the emphasis to the Kuwait field army.

4. Support ground operations.[58]

In reality, the first three phases were executed simultaneously.[59] Air superiority and SEAD, in fact, merged as a function of how the war was fought: the war against the IADS consisted of attacking elements of the Iraqi C^3 structure, its air interceptor force, its early warning/acquisition net, and its radar-directed SAMs all at once, not as abstractions of a written plan that may not have applied. The campaign was not restricted to using electronic combat assets only, but used the entire available array of aircraft and weapons to suppress the totality of the Iraqi IADS, not bits and pieces of it.

Centralized joint command and decentralized execution were the hallmark of the campaign. F-117s were employed in striking key C^3 structures; F-15Es and F-111Fs destroyed key sites in concert with F-4Gs and EF-111s; EC-130Hs jammed critical battlefield communication links with Iraqi fighter aircraft and radio-dependent firing units. What could not be jammed was targeted with ARMs; what could not be attacked by either jamming or ARMs was slated for destruction using standoff ordnance.

The overall SEAD strategy was initially pulled from each of the services. From the Navy, EF-111s employed the concept of limited escort. From the Army, Apache helicopters using infrared missiles were dispatched to destroy selected SAM sites in the Kuwait theater of operations. The Army multiple rocket launch system (MRLS) and other artillery were used to support the SEAD campaign before and during the ground offensive in support of air operations. The Air Force concept of integrated electronic combat was the glue that cemented the campaign. The joint force commander used Air Force electronic combat decision-making concepts and structures in formulating and executing day-to-day JSEAD operations. Navy EA-6Bs and EF-111s were used interchangeably in many operations, and Navy HARM shooters accounted for approximately half of the HARM missiles fired in the war.[60]

SEAD also adhered to the concepts of the airland battle, the three-dimensional "shaping of the battlefield," and the concept of "EC rollback" (electronic combat). By the end of Desert Storm, a clearly definable trend toward adapting both the Army AirLand Battle doctrine and the Navy's philosophy of letting the situation dictate the tactics and weapons selection had resulted in the creation of an integrated, cohesive war-fighting instrument that viewed the total battlefield all at once. Use of this concept would obligate the enemy to behave in predictable patterns and would physically define the dimensions in which the enemy was constrained to operate.

Another feature of SEAD operations in Desert Storm was its adaptation of both Bekaa Valley and Libyan-style operations. The prosecution of the SEAD campaign against Iraq was, in a sense, an extension of the raid mentality of Libya. The application of SEAD during Desert Storm was continuous and unrelenting, it destabilized the IADS and never allowed it to recover. The Libyan operation lasted only 11 minutes; the air operation in Iraq six weeks. But the pressure exerted on the IADS was continuous during both. The Bekaa Valley campaign, likewise, was a model in that it was a closely coordinated, intricately timed operation involving all the services at once in the initial phases of combat. After the first 20 minutes of combat, the Iraqi IADS—even though much of its structure lay intact—ceased to contribute effectively to the Iraqi war effort in much the same way that the Bekaa Valley IADs lay in ruins after the first 20 minutes of air operations.

SEAD had evolved to the point where its concepts and influences became precursors to other aspects of the air power mission. The distinction between command, control, and communications countermeasures, SEAD, offensive/defensive counterair, and battlefield air interdiction appear to have been permanently blurred by the integration of total-force structures in the achievement of total-force objectives. SEAD had started to become JSEAD.

Table 4 implies a correlation between the evolution of SEAD and the convergence of tactics, strategy, and doctrine for the US armed forces. The influences of politics and economics seem to have accelerated rather than moderated the importance of SEAD (and JSEAD) as a major player in future wars.

Table 4

Gulf War Matrix
Operation Desert Storm: Jan–Mar 1991

Threat	Most formidable faced by American military based on capability, numbers, command and control structure, and modernization. Country-wide radar-warning net consists of multiple Western-designed and Soviet radar systems tied together with redundant and hardened command, control, and communications (C^3) net. Advanced computer technology employed in target-tracking, weapons allocation, and sector control. Iraqi air force, AAA, and SAM systems fully integrated. Frontline personnel well trained, secondary back-up echelon poorly trained. Doctrinally tied to "weapons close-hold" use; poor discipline results in random, wasteful use of weapons.
Definition	Campaign-based SEAD. Use of all aircraft to execute SEAD campaign. Difference between way campaign was defined and way it was executed. Definition separated offensive counter air (air-to-air) and SEAD (suppression of radars). Executed as single, coordinated IADS attack plan. Distinction drawn between tactical and strategic SEAD. Tactical SEAD related to day-to-day suppression of battlefield operations against Iraqi field army. Strategic SEAD directed at degradation of overall Iraqi IADS.
Tactics	Full spectrum. Electronic combat-based tactical deception, integrated use of EF-111/EA-6, HARM shooters (F-4G, F-16C, Navy aircraft). Near real-time relay of critical electronic combat data coordinated through AWACS E-3, RC/EC-135, E-2C, prototype JSTARS. Neo-Iron Hand tactics against known sites using combination of jamming ARMs, standoff ordnance (glide bombs, Mavericks, Hellfire missiles, cluster bomb units, or general-purpose bombs).
Organization	Centralized control, decentralized execution. Strategy, general battle plan tactics developed by joint coalition staff. Specific elements of plan tasked to specific units. Developed as war-planning organization as opposed to raid or campaign of Libya or Bekaa Valley operations. Heavily dependent on reliable, efficient communications among forces to make organization viable.
Force Structure	Very heavy emphasis (above 50 percent of force) dedicated to SEAD effort during initial phases. Actual percentage of SEAD-specific assets very small (EF-111s, EA-6s, F-4Gs, EC-130Hs accounted for less than 3 percent of total air forces). Specialization of SEAD assets diminished, emphasis on object of SEAD campaign with whatever assets could perform task.
Strategy	Highly evolved SEAD strategy. Multileveled from strategic to tactical, from specialized function to general campaign. Integrated with doctrine. Plan: destabilize the Iraqi IADS and keep it destabilized. Neutralize those portions that pose a threat to attackers, deny their IADS the use of electromagnetic spectrum, while exploiting other elements of the spectrum for deception and intelligence-gathering purposes.

Table 4 cont'd

	An extended use of Libyan raid mentality. By relentlessly keeping pressure on IADS through jamming, ARMs, selected Iron Hand attacks, seized the initiative by maximizing confusion. Instead of 11 minutes of confusion, there were six weeks.
	The intentional shaping of the electronic battlefield. A use of airland battle concept extended to three dimensions and encompassing the use of the electromagnetic spectrum.
	Variation of Navy's "electronic combat rollback" emphasizing versatile use of aircraft in executing SEAD.
	Based on striking first, prepared to repel attack if it came and retake initiative.
Doctrine	AFM 1-1, 16 Mar 84, still in force. SEAD subsumed as elemet of counter air, and also element of electronic combat. Joint doctrine separated JSEAD from C^3CM, and subsumed electronic warfare under C^3CM. Desert Storm combined elements of both in strategic/tactical applications.
Technology	Very sophisticated technology used by both sides. Technology fairly evenly matched.
	SAMs employed latest Soviet/Western guidance techniques and electronic counter-countermeasure advances.
	Fiber optics, highly directional microwave communication net, state of the art.
	Iraqi fighters practiced fully coordinated night GCI operations using latest Soviet and French fighters and air-to-air missiles.
	Technology too sophisticated to tackle immediately; months-long preparation required, in some cases, to design specific counters.
	US use of digitally reprogrammable equipment critical to success of adapting to rapid changes in Iraqi use of electromagnetic spectrum, especially for HARMs and radar warning receivers.
Political	Strong support throughout. Well-orchestrated coalition to acquire needed authority to use armed force.
	Unspoken objectives politically motivated; minimum US casualties, rapidly moving war with clearly visible, continued successes terminating in unconditional, unambiguous military victory.
	Political objective to "stabilize regional relationship" unclear in pragmatic terms. Possible result: reintroduction of armed force to region at later time.
	National interest defined in economic and altruistic terms: "Free Kuwait" and "Protect the oil resource" used interchangeably.
Economic	Increasingly strong determinant in military/political behavior. Cost of military venture borne by coalition (corollary: destructively expensive to US economy to bear alone).
	Cost of developing, acquiring, and maintaining specialized SEAD mission equipment seen as prohibitive. Advanced upgrades for EF-111, follow-on Wild Weasel cancelled in favor of subsuming SEAD as role of multimission aircraft.
	Only brief hiatus in general economic downward trend. Defense spending under continued fire to reduce in post–cold-war era. Will place increasing pressure on SEAD technology and tactics to "do more with less."

The Larger Context

The matrix shown in table 5 illustrates that SEAD has evolved through use in combat on its own in much the same way as other forms of aerial warfare, except that the evolution has been much faster. In its development, SEAD has grown to encompass or influence many other facets of the air power mission such as strategic bombardment and air superiority. It has had the odd effect of blurring distinctions between previously clear categories.

Table 5

Evolutionary Trends in SEAD: A Brief Summary

Threat	Movement from preeminence of strategic power over defense to credible use of air deniability air defense systems to balance between strategic offensive and IADS defensive; return to preeminence of strategic/tactical air power of air defense (if tactics properly executed).
Definitions	Moving from specific applications (defense against SAMs) to more generalized applications of neutralizing an enemy's ability to defend from an air attack.
	Moving from purely tactical uses of SEAD to a mix of strategic and tactical uses to (1) neutralize the IADS for strategic bombing and (2) protect friendly aircraft.
	Moving from a purely support role to a campaign objective in direct proportion to growth of "air deniability" strategies and IADS/organic air defense capabilities.
Tactics	Movement from brute-force attacks with no specialized equipment to use of specialized-only assets, to mixture of specialized defensive/offensive tactics.
Organizations	Moving from no organization at all (prior to Vietnam) to very specific organization (66th ECW, Twelfth AF EC assets) to SEAD organizations diffused throughout the TAF that are mission based.
	Movement to complex, multistranded, confusing organization; to simplified, specialized organization; to complex, integrated, streamlined organization.
Force Structure	Movement from none to limited specialized, mixed general-purpose forces to specialized, highly defined forces to specialized mix of special-purpose aircraft and mixed joint-force assets.
Strategy	Movement away from strategic nuclear doctrine to flexibility of air power applications in rapidly changing world. Responsiveness, versatility, flexibility will be key components of future SEAD strategy.
	Changes in economics forcing less sophisticated applications of SEAD. Movement away from specialization and toward pooled joint resource applications of existing technologies to accomplish SEAD functions.
Doctrine	Movement from specific area/ground-based applications for defending aircraft to major element of counter air to campaign objective.

Table 5 cont'd

	Probable movement to encompass Navy concept of multiple-role SEAD (ARM) and application of three-dimensional shaping of electronic battlefield.
Technology	Exponential growth in quality/quantity. Rapid changes make whole families of equipment/tactics obsolete. Accelerated change has dramatic impact on nature of tactics and force structure. Overall strategy lags tactics as function of technology. Doctrine lags strategy. By time of Desert Storm, strategy and doctrine driven by tactics derived to defeat specific technologies. Impact of economic drawdown will have general stabilizing effect on technology. Result: technology will have less influence on tactics, hence strategy and doctrine become preeminent.
Political	Movement from bipolar-influenced, regional conflicts to regional conflicts where superpowers exerted secondary influence (Bekaa Valley) to superpower/third world confrontation tenuous to negative support (Libya) to coalition, multipolar, regional conflicts.
	Economic prerogatives becoming more determinative in international behavior. National actions exert increasing pressure of politico-military affairs.
Economic	Movement from relatively inexpensive equipment (quantity) to large amounts of relatively specialized equipment (much of it rendered obsolete by rapid technological change) to smaller and smaller amounts of extremely expensive, adaptive equipment to the creation of a mixture of some expensive equipment and inexpensive equipment because of extreme costs of higher technologies. Cost no longer justifiable in terms of anticipated threat.
	Some aspects of SEAD to become generalized function of multirole aircraft (F-16, F-22) in order to save money. Optimum solution no longer tenable; minimally acceptable solution to SEAD problem more and more likely due to fiscal constraints.

For instance, the destruction of the Iraqi strategic IADS and C^3 net was a function of air superiority, strategic bombardment, and SEAD all at the same time. In a sense, the fabled "MiG-sweep" concepts of the Korean and Vietnam wars have been replaced by the more encompassing "IADS sweeps" of Desert Storm. It would be shortsighted to point to any one variable in the matrix as the primary determinant, but certainly technology stands out as a key factor in forcing the evolution of SEAD and in propelling formerly separate categories and missions toward a more integrated, convergent sum of concepts now assumed under JSEAD.

In the larger context, as the world becomes increasingly complex, as military strategists and international relations theorists struggle to create paradigms that define the "new world order," a real US military must live and operate in the real world. Whatever role is crafted for the US military in the new world order, US forces' basic traditional role in fighting and winning wars is not likely to be changed for some time. Nonetheless, there is a clear—and troubling—shift from a military which *defends America* to one which *defends the national interest.* The former is much clearer, the latter much less so. The "national interest" is dependent on definitions that are no longer clear.

Clarity in defending the national interest has direct relevance to joint war fighting in the future and to how joint SEAD must be configured to be meaningful in the execution of military strategy. That is why an understanding of SEAD's evolution is so critical to air power application.

Notes

1. Dennis M. Drew and Donald M. Snow, *Making Strategy: An Introduction to National Security Processes and Problems* (Maxwell AFB, Ala.: Air University Press, 1988), 163. Italics in original.

2. William P. Snyder, *Military Studies Course, Book 1* (Maxwell AFB, Ala.: Air War College, 1991), 49–55.

3. Mark Clodfelter, *The Limits of Air Power: The American Bombing of North Vietnam* (New York: The Free Press, 1989), 203–10.

4. Earl H. Tilford, Jr., *Setup: What the Air Force Did in Vietnam and Why* (Maxwell AFB, Ala.: Air University Press, 1991), 89–95.

5. Ibid., 122–23.

6. Ibid., 125–27.

7. Ibid., 124.

8. Ibid., 241.

9. Robert Frank Futrell, *Ideas, Concepts, Doctrine: Basic Thinking in the United States Air Force, 1961–1984,* vol. 2 (Maxwell AFB, Ala.: Air University Press, 1989), 291.

10. Mario de Arcangelis, *Electronic Warfare: From the Battle of Tushima to the Falklands and Lebanon Conflicts* (Dorset, UK: Blandford Press, 1985), 168–70 (translated from the Italian). Arcangelis' comments give an apt description of hunter-killer tactics and their predecessor, Iron Hand sorties. Hunter-killer tactics were characterized by two Wild Weasels working in concert: one would stimulate the site and keep it on the air while the other acquired the signal, identified it, located it, and fired either an AGM-45 or, later in the war, the AGM-78 Standard ARM. The Iron Hand strikes were usually characterized by direct attacks on the SAM sites with iron bombs and little or no Weasel support. The combination of the two proved very effective in the latter stages of the war. The situation posed by combining tactics of both hunter-killer and Iron Hand teams resulted in a "lose-lose" proposition for the SAM site operator. If he turned on his radar, the hunter-killer ARMs would target him; if he turned off his radar, the Iron Hand bombers would destroy him since, out of fear of being targeted himself, he would have decided not to use his radar to defend against the attack.

11. Tilford, 240.
12. Ibid.
13. Ibid., 241.
14. Ibid., 255.
15. Ibid.
16. *The Battle for the Skies over North Vietnam* (Maxwell AFB, Ala.: Air Command and Staff College, 1978), 95.
17. Tilford, 257.
18. Ze'ev Schiff, *The History of the Israeli Army 1974 to the Present* (New York: Macmillan, 1985), 239–45.
19. Futrell, 556.
20. Arcangelis, 190–91.
21. Schiff, 217.
22. Paul S. Cutter, ed. "ELTA Plays a Decisive Role in the EOB Scenario," *Military Electronics/Countermeasures*, January 1983, 136.
23. Lt Gen Kelley H. Burke, USAF, Retired, "Electronic Combat: Warfare of the Future," *Jewish Institute for National Security Affairs Newsletter*, February 1983.
24. Kenneth P. Werrell, *Archie, Flak, AAA, and SAM: A Short Operational History of Ground-Based Air Defense* (Maxwell AFB, Ala.: Air University Press, 1988), 140–42. The Syrian ZSU-23-4 AAA weapon employed by the Syrians enjoyed a good deal of exposure, and a considerable body of apocrypha has grown up around it. A favorite story told by Israeli pilots is that, on a road reconnaissance mission, a formation of four Israeli fighters flew over a ridge line to discover a column of Syrian troops protected by a ZSU-23-4. They immediately maneuvered to avoid the system, but in the pull (a period of 15–25 seconds, so the story goes) three fighters were shot down outright, and the fourth badly damaged.
25. Ronald T. Pretty, ed., *Jane's Weapon Systems, 1982–83* (London: Jane's Publishing Company, 1982), 480–87.
26. Chris Heath, "Electronic Warfare: The Lessons of 1982," *Pacific Defence Reporter*, December 1982/January 1983, 68.
27. John V. Cignatta, "A U.S. Pilot Looks at the Order of Battle, Bekaa Valley Operations," *Military Electronics/Countermeasures* 9, no. 2 (February 1983): 107.
28. Schiff, 245.
29. Ibid.
30. David Clary, "EW in the Bekaa Valley: A New Look," *Journal of Electronic Defense*, June 1990, 38–39.
31. Paul S. Cutter, "Lt. Gen. Rafael Eitan: 'We Learned Both Tactical and Technical Lessons in Lebanon,' " *Military Electronics/Countermeasures* 9, no. 2 (February 1983): 94.
32. David Clary, *The Bekaa Valley: A Case Study* (Maxwell AFB, Ala.: Air Command and Staff College, 1988), 12–14.
33. Charles E. Mayo, "Lebanon: An Air Defense Analysis, *Air Defense Artillery*, Winter 1983, 22; and Philip J. Mills, "RPVs over the Bekaa Valley," *Army*, June 1983, 49. It is difficult to determine the exact types and mixes of weapons from the record. Some note that there may have been as few as 15 SA-6 sites (Mayo), others indicate a number greater than 20. The literature,

of which Mayo and Mills are part, spans total numbers of systems from a low of 15 total sites to as high as 31 over a period of six weeks.

34. Arcangelis, 190.

35. Clary, *The Bekaa Valley*, 11–14.

36. Clary, "EW in the Bekaa Valley," 38.

37. Clary, *The Bekaa Valley*, 15.

38. "IAF vs SAM: 28:0," *Defence Update International* 78, no. 12 (December 1986): 54. Again, the reader may note that the number of SAM sites this article cites is 28 as opposed to 19 or 31. The source *Defence Update International* cites in support was a Soviet defense analyst, a Col V. Dubrov.

39. Mayo, 24.

40. Schiff, 250.

41. Clary, *The Bekaa Valley*, 11.

42. "Modern War Wizards," *Defence Update International* 78, no. 12 (December 1986): 45–50.

43. The proliferation of the use of computer-aided decision-making devices by the mid-1980s was widespread among battle-planning staffs. The electronic combat community, being more technically oriented than many, was quick to exploit the systems available. This enabled types of planning undreamed of before. Some of these computers could very accurately depict an enemy radar's detection capabilities and overlay US jamming aircraft's capabilities to degrade it. Using the computers also saved many man-hours of work formerly accomplished with a grease pencil on a map that was manually updated with information of unknown origin and unknown timeliness.

44. Frank Elliot and Len Famiglietti, "Planning, Precision Make Libya Raid Succeed," *Air Force Times*, 28 April 1986.

45. The trade journals openly advertised their sales to foreign nations, and Libya was a buyer. The Libyans apparently were able to mate dissimilar systems into a more or less coherent net, though it is not understood how well this system worked.

46. The tactics used against specific sites resembled a combination of Iron Hand and hunter-killer methods. The tactics appeared also to have been altered by the technologies available at the time.

47. Elliot and Famiglietti, 30.

48. "US Airpower Hits Back," *Defence Update International* 73, no. 8 (August 1986): 27–33.

49. John W. R. Taylor, ed., *Jane's All the World's Aircraft, 1987–88* (London: Jane's Publishing Company, 1987), 428–29; and *Jane's All the World's Aircraft, 1982–83*, 380.

50. Interviews with United States Air Force Europe El Dorado Canyon participants, Headquarters, Tactical Air Command, Langley AFB, Va., 1 August 1991.

51. Interviews with US Navy aviator participants of El Dorado Canyon, Headquarters, Pacific Air Forces, Hickam AFB, Hawaii, March 1991.

52. Fact-finding trip, Headquarters, United States Air Force, Washington, D.C., 28 July 1991. During a visit to HQ USAF/SA (the former title for Studies and Analysis), I discovered that most of the analysis being

performed was involved in processing and cataloging the mountains of data that was being collected on Desert Storm.

53. James Blackwell, *Thunder in the Desert* (New York: Bantam Books, 1991), 96; and Norman Friedman, *Desert Victory: The War for Kuwait* (Annapolis, Md.: Naval Institute Press, 1991), 74–108.

54. "After the Storm," *Jane's Defence Weekly*, 6 April 1991, 529–31.

55. *International Air Forces and Military Aircraft Directory* (Essex, UK: Aviation Advisory Service, July 1991), 209–10; and *Jane's All the World's Aircraft, 1991–92*, 145.

56. Bert Kinzey, *The Fury of Desert Storm: The Air Campaign* (Blue Ridge Summit, Pa.: TAB Books, 1991), 15.

57. Interviews with USAF participants in Desert Storm, Maxwell AFB, Ala., August 1991.

58. Kinzey, 12.

59. Col John Warden, USAF, Headquarters, USAF, Washington, D.C., interview with author 30 July 1991.

60. Kinzey, 120.

Chapter 2

Criteria for Assessing SEAD Effectiveness

Few subjects in electronic combat (EC) frustrate tacticians, as much as that of developing meaningful criteria to assess electronic combat effectiveness. As an integral part of EC, suppression of enemy air defenses (SEAD) also has defied attempts to measure its effectiveness. Most strategists and tacticians assent to the great value of SEAD as an element of modern warfare. Though strategists agree that SEAD will increase in importance as a war-fighting tool, few of them agree on how to assess its value.

This chapter outlines traditional models used in shaping criteria as a precursor to developing criteria that better assess SEAD's effectiveness. The latter portion of this chapter develops some contextual themes that provide a bridge between the traditional models used today and some suggested ways of assessing SEAD's value in the uncertain world of tomorrow.

Suppression of enemy air defenses—as an agglomeration of high-technology weapons and apparatus, old and new tactics, and traditional forces and concepts—engenders the same paradoxes that befell the tank, the airplane, the submarine of World War I, and nuclear weapons of World War II. Modern SEAD weapons and concepts arrived on the scene long before holistic patterns of use were developed for them. The impetus to develop SEAD weapons unwittingly brought about war-fighting technologies that had unforeseen consequences on the modern battlefield. SEAD weapons and tactics accelerated modern warfare's evolutionary process so much that traditional methods of determining effectiveness for SEAD no longer make much sense. SEAD evolved from the Iron Hand tactics of Vietnam to the sophisticated multiphased, multispectrum SEAD tactics used in Desert Storm in the short space of 27 years. SEAD effectiveness in Vietnam was measured generally by counting destroyed SAM sites and radars. Applying that criteria to the SEAD technologies used in

Desert Storm yields a confused, possibly irrelevant picture. SEAD weapons and tactics evolution has outpaced the development of criteria to measure SEAD's total contribution to combat.

War effects traditionally have been fairly easy to discern: so many tanks destroyed; so many troops killed, wounded, or missing; so many bombs dropped; so much space occupied by friendly troops. These are tangible, measurable effects. When effects yield easily to quantifiable, observable means of analysis, developing meaningful methods of assessing effectiveness is straightforward. One can measure how far away a bomb hit from the intended target to determine accuracy. One can assess number of hits per try—or number of targets destroyed per mission—to develop understandable criteria for effectiveness.

Unfortunately this is not true for SEAD. Since the electromagnetic spectrum is the medium with which modern defense suppression works, the criterion of counting bomb craters, measuring distances from intended impact point, or assessing hits per try is more difficult to apply. Wave fronts of electromagnetic radiation are invisible; the theory of their propagation is not easy to explain—nor is it easy to counter an adversary's use of the spectrum. Electronic combat literature is replete with references to this difficulty, and several approaches have been offered. Most of these approaches, however, still attempt to assess SEAD effectiveness as a function of observable phenomena. Many tactics and strategy analysts try to place SEAD effects in the realm of the observable without clear reference to the overall objective SEAD supports. This assumption, while understandable, can result in unrealistic appraisals of SEAD effectiveness. For instance, a traditional measure of effectiveness is probability of kill (P_k). This measure is a combination of historical and engineering test data that tells the user what the probability is of destroying a specific site under specific conditions. Let us assume an antiradiation missile has a P_k of .4 against a specific enemy system. If one wanted to guarantee a kill on the enemy system of .9, at least three missiles would have to be employed against each site. To arrive at the total number of missiles required for theater operations, one determines the P_k desired and the number of sites slated for destruction via antiradiation missiles (ARMs), and performs the appropriate mathematical calculations. Employing this empirical-appearing process seems to be completely rational; however, it ignores the specific objectives a battlefield commander may have, the situations in which the antiradiation missiles systems may be employed as part of a larger plan, the capabilities of other SEAD systems at the commander's disposal, or ancillary effects that ARMs may have on the enemy that outweigh system P_k (i.e., psychological warfare against enemy site operators via intimidation).

Methods of Analysis

In a medium where the effects are not easily observable, often not directed at destruction, nor related to factors given to quantifiable analysis, the attempt to force a straightforward "hits-per-try" analysis model may fall short. The SEAD quandary is a classic case of trying to fit a square peg (SEAD) into a round hole (counting bombs and bullets). Nonetheless, previous approaches that have grappled with assessing SEAD effectiveness provide an excellent place to start. These viewpoints represent a broad spectrum of thought, and many are not even "formal" methods of analysis.

We will place these approaches into four categories: (1) the historical model, (2) the engineering model, (3) the commonsense model, and (4) the objective-based model. These models represent, like the trends presented in chapter 1, an evolutionary pattern of development. Each successive method of evaluating SEAD's effectiveness is the result of multiple historical factors propelled primarily by technological developments. The evolution of measurements of SEAD effectiveness and the criteria that generate these measures are interrelated and complex. Each successive model builds on its predecessor in the attempt to relate a meaningful criteria to a relevant, usable measure of SEAD effectiveness.

The historical, engineering, and commonsense models represent the evolutionary process to the present. While elements of each exist in the attempt to measure SEAD effectiveness, there are traceable trends in the relative blend of these approaches over time. The shock of winning the cold war so suddenly—with the resultant paradox of having relatively less SEAD capability to fight an increasingly sophisticated future adversary—creates a distinct watershed for SEAD. It no longer makes sense to use measures of SEAD effectiveness and criteria meant to assess the military worth of strategies, doctrine, weapons, and tactics during a cold war that no longer exists.

Providing sufficient SEAD capability for future conflicts—and a means of determining what SEAD options are more effective than others—will be problematic. A new model—the objective-based model—is emerging to help solve this dilemma. The objective-based model is both a combination of the other three and a bridge into the future. It is an attempt to assess where measurements of SEAD effectiveness are going as a result of evolutionary trends. The objective-based model offers an alternative view of the traditional criteria-building process by adapting a popular political science model, a variation of Karl Deutsch's "cybernetic" process described in his book *The Nerves of Government.*[1] In very general terms, the cybernetic process refers to self-adjusting mechanisms that respond to multiple factors in an environment to achieve equilibrium. The thermostat on an air conditioner is a simple example.

An objective-based criterion for measuring SEAD effectiveness is based on a process which relates SEAD employment directly to overall objectives. The objective becomes the driving element in achieving equilibrium, and all other factors in the equilibrium equation are related to the objective.

The objective-based criterion, as the product of evolution, also provides a framework in which future SEAD weapons and tactics may be set into an overall context. In setting the context, four continuums combine to explain the general trends facing the development of objective-based criteria. The continuums provide a context, a way of determining what SEAD strategies, tactics, doctrines, weapons acquisitions and resource allocation plans will be more effective. More importantly, because of the speed of the evolutionary process, the continuums provide a mental construct with which to anticipate future SEAD needs and changes. These continuums define a range of options as well as depict an evolutionary trend in SEAD. These are

1. the piecemeal/integrated continuum,
2. the need-based/resource-based continuum,
3. the threat-based/capability-based continuum, and
4. the defensive/offensive continuum.

The continuums provide a way of assessing SEAD effectiveness by describing the changed context of the world. They describe an evolutionary process that provides a way of comparing one criterion with another and shows that criteria development itself is subject to the same forces of disequilibrium that affect the revolutionary growth of SEAD.

The Historical Approach

The historical approach is usually the first approach decision makers turn to when confronted with the problem of criteria development and measures of effectiveness (MOE). This approach suggests that the lessons of history can serve as a basis for viewing the future and can provide a model of assessment. The initial use of fighter-bombers for direct attack in suppressing surface-to-air missile (SAM) sites during the early phases of the Vietnam War is an example of this approach and the conditions under which this approach might be used. There was no systemic analysis conducted, no elaborate mathematical model used in developing this relatively crude SEAD tactic. This Iron Hand defense suppression tactic was an outgrowth of World War II and Korean War tactics to silence antiaircraft artillery sites; the Iron Hand tactic was used because similar tactics were effective in the European theater of World War II and in Korea.

The recourse to historical comparison was the initial response to developing a SEAD criteria. While the direct-attack method worked in World War II and Korea, it soon became obvious that this particular historical approach would not work in Vietnam. Exposure to the Soviet SAM in Vietnam was the first experience American tacticians had had with a relatively sophisticated technological threat. Recourse to historical comparison, while useful in conditions where technological breakthroughs were not critical to war-fighting outcomes, rapidly proved to be of limited value as a source of SEAD criteria in Vietnam.

Chapter 1 of this study could also be viewed as an example of the historical approach. In viewing "what worked before," however, one must be wary of the assumptions used in recounting the history itself. The intent of chapter 1 was to show the evolutionary nature of SEAD combat by tracing the main elements of its growth and scope, not to establish history as a model of the future.

There are problems in using the historical model as a primary vehicle in developing criteria for measuring SEAD effectiveness. SEAD, as a technologically driven aspect of war fighting, requires a certain amount of prediction for its effective use. History does not seek to predict; it intends to recount facts and explain why an event occurred as a function of hindsight. SEAD application is often quite technical and needs data on what will work *next* as much as it needs data on why a particular weapon or tactic did or did not work *last time.* As a tool in proactive analysis—perhaps even as a basis for creating a criterion for comparing one period of time to another—the historical approach is of limited use.

Examples of the *mis*application of the historical approach are revealing. Using the historical approach, a criterion for SEAD effectiveness in Desert Storm might be to compare the loss rates of Desert Storm aircraft which flew across enemy radar-directed SAM complexes to similar missions in Vietnam. But recounting the pertinent events of these periods only explains what happened, not why, and offers no substantive construct upon which a criterion for evaluating effectiveness can be built. The historical comparison reveals another weakness: one period of history cannot be compared directly to another unless all the pertinent variables are held relatively constant. As Karl Popper points out in his book, *The Poverty of Historicism,* "For strictly logical reasons, it is impossible for us to predict the course of future history."[2]

Using another avenue of comparison, the analyst may wish to compare EB-66 jamming in Vietnam to EF-111/EA-6 jamming in Desert Storm. The record suggests, based on loss rates, that EB-66 jamming in Vietnam was not as effective, but this does not necessarily mean that the tactics employed in

Desert Storm would have worked better in Vietnam, or that Desert Storm SEAD tactics will work as well "the next time." Therefore, a straightforward comparison of the historical record, from the criteria development vantage point, can be misleading.

It would be imprudent to discount the historical method altogether because it can offer compelling arguments as to why one course of action might be pursued over another. As pointed out, some of the lessons learned in the past and apparently applied during the Desert Storm SEAD campaign seem to have enabled friendly force planners and tacticians (1) to use jammers in places where they would do the most good at the times they would do the most good; (2) to identify what could and could not be done with the jamming aircraft; and, (3) to balance tactics and force application with sister "lethal" SEAD assets (Wild Weasels and Iron Hand bombers).

The unspoken assumption, nonetheless, is that the historical "facts" recounted represent a "truth," and that the assumptions surrounding the selection of similar situations for comparison can stand on their own without scrutiny. To criteria developers who use the historical model as their primary determinant, one should attach a "user-beware" notice to match the "right" historical model with the "right" situations. This will become especially important for the development of SEAD criteria in the future.

Another outgrowth of the historical method is the "lessons-learned" approach. A great deal of emphasis has been placed on this approach as a result of Desert Storm; it is a classic example of what Popper would call "the poverty of historicism": we have no way of knowing whether the lessons learned will apply in the future because we have no way of knowing the future.[3] The danger of the "lessons-learned syndrome" is that it focuses attention on the events of the past as a function of hindsight, not on the possibilities of the future as a function of foresight. No matter how hard we attempt to project ourselves into the future and no matter how accurate the data we are given in the attempt to project history into the future, we must use today's paradigm in the projection. Even if the conditions of the "real world" remain relatively stable (an unlikely situation), the peculiarities of bureaucracy, force structure and organization, the personalities of leaders, and the reordering of priorities caused by purely nonmilitary factors make it improbable that these "lessons learned" will provide the basis for developing sound criteria.

While "lessons learned" are doubtless valuable, they have an irritating tendency to be forgotten. The "rediscovery" of the need for on-the-spot suppression of battlefield SAMs and AAA was a recurrent theme for the close air support (CAS) and battlefield air interdiction (BAI) roles flown by F-16s, A-6s, A-10s, and F/A-18s. Yet the after-action reports and "lessons learned" from major exercises held over the last 10 years highlighted the need for integral battlefield SEAD. The more important question in this regard is not what the "lessons learned" were, but why warriors have to keep relearning them. The answer is found in the nature of the historical model itself. The historical approach provides direction and trend, but not necessarily immutable truths to which we might apply criteria of measurement.

The Engineering Approach

The engineering approach is characterized by the attempt to measure, precisely, what can be measured. The approach is influenced strongly by scientific methodology and technological innovation. The clearest examples of the engineering approach manifest themselves in systems analysis and the compulsion to reduce information to quantifiable data. The process is, like modern SEAD, highly technical and complex. The complexity of the engineering approach often obscures the objectives for which SEAD exists with the technical process of assessment. The principal problem is that the engineering approach does not seek to relate conclusions to overall SEAD objectives nor does it ask questions on its own. The engineering approach is designed to answer questions that are posed to it from people outside the assessment process. Actors exterior to the process give SEAD engineers specific tasks—for example, determine the effectiveness of this weapon system—and, from the context of the given objective, engineers create a methodology for answering the critical question, "Exactly what is it that you want to know about this?" Responsibility for ensuring that the overall objective is germane to the general SEAD problem is rarely a major issue; consequently, many measurements of effectiveness assess variables which mean very little in terms of overall objectives. For instance, studies of the effectiveness of the AGM-88 antiradiation missile (the HARM), which tracks radar emissions of hostile radars to their source, take into account many of the pertinent variables (ballistic flight path of the missile; angles and speed of flight; time required to acquire, identify, and track a target; parametric data of the intended target; etc.). The results of such studies (both before and after conflicts) yield probabilities and other statistical measurements. In the parlance of the tactics community, one wishes to arrive at a credible probability of kill figure, which, by inference, would reveal how "effective" the weapon is. As accurate as these models are, specific system P_k will explain only part of the puzzle. Just because such a study examines data which is quantifiable and represents discretely defined variables does not mean it has examined *all* the critical determinants bearing on the problem. Determining *real* effectiveness is more complex than this.

The engineering approach led to a practice quite common in the late 1970s/early 1980s: the substitution of efficiency for effectiveness as the criteria for establishing military worth. This led to an acquisition and testing process that often divorced the purpose for which a SEAD weapon was developed from the measuring stick used to determine its worth. As long as the weapon met the efficiency standard set by the engineering community, decision makers judged it a worthy system.

Herein lies a key danger in applying the engineering approach as the principal source of SEAD criteria selection: it is not designed to build an overall framework that relates objectives to methods of assessment. As a result, the outputs are only as good as the assumptions going into the model and the relevance of the variables selected for examination.

As the engineering approach is the most dominant method used by the military in determining measures of SEAD effectiveness and criteria formation, it is worth scrutinizing closely. We will use the Air Force's problem-ridden ALQ-165 advanced self-protection jammer (ASPJ) program as an example of the engineering approach in criteria selection.

The story of the ALQ-165 begins with an examination of the definitions used by the engineering community. Such an examination reveals the complexities and difficulties one encounters in employing the engineering approach as a primary determinant in criteria selection. What do we mean by "criteria" and "measure of effectiveness"? "Criteria," oddly, is not defined by itself in the military lexicon. It is identified, however, in association with "evaluation criteria" and defined as

> standards by which required technical and operational effectiveness and suitability characteristics or the resolution of technical or operational issues may be addressed.[4]

This definition is very narrow in that it specifies "technical and operational effectiveness and suitability" as critical items. Operational effectiveness and suitability are further defined and subdivided in a laundry list of smaller and smaller facets. This has the ironic effect of making the job of quantification less and less credible as the respective elements become more and more vague. Operational effectiveness and suitability are defined, respectively, as

> the overall degree of mission accomplishment of a system when used by representative personnel in the environment planned or expected for operational employment of the system considering organization, doctrine, tactics, survivability, vulnerability, and threat (including countermeasures and nuclear threats);[5]

and

> the degree to which a system can be placed satisfactorily in field use, with consideration given to availability, compatibility, transportability, interoperability, reliability, wartime usage rates, maintainability, safety, human factors, manpower supportability, logistic supportability including software supportability, documentation, and training requirements.[6]

It is not surprising, based on the complexities involved in official definitions, that *no specific criteria* were ever set down by the USAF Tactical Air Warfare Center or the Air Force Electronic Warfare Center for developing an ASPJ measure of effectiveness (MOE). In fact, one of the criticisms of the process by the Tactical Air Warfare Center was that "selecting an MOE solves only a portion of the problem. A criterion for that MOE is required. . . ."[7]

The word *criterion,* as defined in a standard dictionary, is broad and captures the more essential elements of the development process: a standard, rule, or test on which a judgement or decision can be based. Rather than focusing on the elements of "technical and operational effectiveness and suitability," this definition implicates the decision maker as part of the process and provides a broader plane of application via the use of rules and tests.

A measure of effectiveness is the functional element of applying criteria. It denominates the standards, rules, or tests in quantifiable, measurable terms. For the ASPJ program, these were manifested as (1) projectile miss distance generated by jamming, (2) hits per projectiles shot, or (3) number of times projectile failed to fire due to jamming per times jamming present. Even if the system performed within the specified limits defined by the criteria, there was no way to determine whether the criteria *upon which the measurements were based* were relevant. The requirement to establish quantifiable and technically measurable variables of the ALQ-165 was so compelling that the focus of the criteria was on the process of measuring rather than on the objective for which the self-protection system was developed.

The regulations governing the MOE process and definitions also reflect the engineering approach's preoccupation with quantifiable criteria. The definition of MOE in AFR 55-43, *Management of Operational Test and Evaluation,* is "a qualitative or quantitative measure of a system's performance, or a characteristic that indicates the degree to which it performs the task or meets a requirement under specified conditions."[8] The definition focuses on specifics: a system's performance or a characteristic. The definition also indicates that the measure specifies the degree to which systems or characteristics perform a task or meet a requirement without identifying who stipulates the requirement, why they want the system to perform the task, or what overall objective is met.

The gulf between decision maker and engineer that is created by this methodology in the development of relevant SEAD criteria is as much the result of the scientific mystique as it is the failure of users and testers to ensure the relevance of their work to overall objectives. It is relatively easy in technocratic culture to "sell" results couched in quantifiable terms. The "experts" are the builders of the graphs and charts and are generally well versed in scientific research/test methodologies. Conversely, decision makers are versed in the overall business of managing competing objectives, bureaucracies, and generalized problem solving. Generally, they are not versed in research methodologies, preferring instead to hear the bottom line. Unfortunately, the bottom line as understood by the technician often is considerably different than that understood by the decision maker. What often happens in the interchange between the decision-making and the engineering bureaucracies is a failure on both sides to communicate the assumptions one community has of the other. The decision maker often assumes the tester knows how the specific criteria fit into an overall scheme of things, and the tester assumes the decision maker knows the limits of the developed criteria in determining overall effectiveness. The result is that decision makers often fail to define the framework into which a specific criterion for testing SEAD effectiveness fits, and the analysts—*because it is not in their job description to set overall objectives—do not challenge the assumptions upon which the tests are based.*

At the more technical level, the specific measures themselves belie their narrowness. The basic formula upon which comparative engineering testing is based is very simple when written out in plain English but has an ominous appearance as a mathematical formula. In plain English it is:

SEAD effectiveness equals the difference between what happens when one uses a particular SEAD weapon/tactic in combat condition as compared to what happens when one doesn't. In mathematical terms, this equation appears as

$$E = P_0 - P_1$$

where P_0 is the probability (expressed as percentage) of survival in conditions where a system or tactic was not used ("dry"). P_1 is the probability of survival where the system or tactic was used ("wet")—all other pertinent and controllable variables being held constant. The difference, E, is expressed as a measure of how much more or less effective the weapon/tactic was in the survival of the aircraft.

The measurement of effectiveness in this case is the assessment of the "probability of survival" in given situations as a function of ASPJ jamming being present or not. The salient point here is that *the measurement is only as good as the assumptions which underlie it,* and these are wanting in many cases.

The narrow mathematical version of the MOE accounts for variables only insofar as they can be quantified and only insofar as they relate to the principal measure of probability of survival. The effects of nonquantifiable variables can have a decisive impact on SEAD measures of effectiveness. Therefore, the mathematical interpretation of effectiveness—as authoritative as it may appear—may be very misleading.

The ALQ-165 ASPJ program serves as an example of what can go wrong with using the engineering approach as the primary determinant in criteria selection. The SEAD effectiveness of the system was related to very specific tasks given to the engineering community. The requirements levied on the system were never challenged by the engineering community, nor were the standards of measuring effectiveness adequately scrutinized by decision makers. The questions relating total system effectiveness—its interoperability with the systems in which the ASPJ was to be used, its viability in the real world, and the political acceptability of its price tag—were addressed *after* the engineering approach had completed its SEAD effectiveness assessment of the ALQ-165, not as part of the assessment process.

Both decision makers and the engineering community are aware of the problems that separate them and are making attempts to rectify the situation. The chart (fig. 5) shows the traditional relationship between the engineering and decision-making community.[9] The hierarchical, linear structure depicted shows the difficulty of translating decision makers' objectives into testable criteria beyond a certain level.

National security objectives are translated to national military objectives. These objectives are, in turn, translated to regional and theater objectives. There is a disconnect in the process that occurs between objectives and operational art. It is not clear how a national military objective supporting national security policy can be translated into a criterion for

AN EVALUATION FRAMEWORK
(WHERE TO TEST AND EVALUATE?)

(POLITICAL/ECONOMIC/ MILITARY)	NATIONAL SECURITY OBJECTIVES
------------- (MILITARY COMPONENT) -------------	
(GLOBAL LEVEL)	NATIONAL MILITARY OBJECTIVES
(THEATER LEVEL)	BROAD REGIONAL OBJECTIVES
(OPERATIONAL LEVEL)	REGIONAL OPERATIONAL OBJECTIVES
	DISCONNECT
(ENGAGEMENT LEVEL)	TASKS — *THE LEVEL FOR IDENTIFYING MISSION DEFICIENCIES*
------------- (HARDWARE SOLUTION) -------------	
(ENGINEERING LEVEL)	FUNCTIONS
(SYSTEMS DESIGN)	SYSTEM CHARACTERISTICS
(COMPONENT LEVEL)	COMPONENT DESIGN

Disconnect between national objectives and tactical objective at operational level.

Figure 5. Hierarchical/linear decision-making model.

assessing the military worth of SEAD, especially in uncertain times.

The measures of SEAD effectiveness have no meaning if they are not tied to military operations, yet the engineering approach does not address these issues. It can aid the process only at the most fundamental level (engagement level). As long as decision makers understand the limits of the engineering approach in answering SEAD effectiveness questions, the engineering approach can be a useful decision-making aid. As long as engineers understand the context of the objective-building process, they can construct tests that best answer the questions decision makers ask.

Dialogue between decision maker/tasker and the engineering community are an encouraging step toward establishing a more encompassing method of criteria development and measure of effectiveness. The dialogue promises to define the limits of what each community can do for the other. Hopefully, this will reduce the gulf which separates the two by demystifying the engineering approach, and by bringing greater incisiveness to defining overall objectives.

Engineering measures serve to explain observable variables, but do not generally provide the basis in and of themselves for criteria selection—especially if the critical variables are not easily translatable to "objective" observation and quantification. In combination with other approaches, the engineering approach serves a very useful function, but it should not—because of its servant nature—be drafted as the lead element in developing criteria.

The Commonsense Approach

SEAD criteria must be able to pass the "commonsense" test. Yet, while SEAD criteria must be understandable, there is an insidious danger in using common sense as the basis of SEAD criteria development. SEAD, while serving straightforward combat objectives, is not simple. No amount of attempting to "simplify" the inherently complex processes in modern SEAD will make it any less complex. Beware the man who appeals to commonsense as his sole reference of authority, wrote nineteenth century American sociologist William Sumner.[10] Common sense is often neither "common," nor does it necessarily represent "sense." Criteria selected from among the realm of "commonsense" elements suffer many ambiguities and errors in logic. First, the commonsense approach suggests a central frame of reference that is assumed to be universally understood and applicable. Second, it fails to account for perceptual differences and assumes that there is only one credible paradigm with which to view the world.

The principal attractive element of the commonsense approach in selecting SEAD criteria is its appeal to simplicity. The commonsense approach for SEAD criteria came into vogue largely as a reaction to perceived shortcomings in the engineering and historical approaches and the belief that these approaches had needlessly complicated the assessment process. The historical and engineering approaches had not successfully delivered systems or tactics that meshed with the perceived needs of the military in an increasingly austere political and fiscal environment.

Many different SEAD and EC systems of the 1980s suffered from the perceived failure of the historical and engineering approaches' measurements of effectiveness. The failure of ASPJ to meet "spec" integration problems with the electronic defensive suite of the B-1, quality control problems with the Tacit Rainbow SEAD harassment drone, and numerous cost overruns throughout SEAD-associated weapons development and acquisition programs cause considerable frustration among the decision-making bureaucracy. In the late 1980s, therefore, there was a general call for a return to "old-fashioned common sense."

On the surface, the "return-to-commonsense" argument has the feel of soundness. In reality, it is very dangerous. The commonsense approach assumes that any complex problem can be stated in simple terms. For SEAD, this is not the case. Simplification of the technological and financial complexity of modern SEAD can be taken only so far. The commonsense approach tends to develop criteria and measure effectiveness on the basis of logic that is generally very superficial. While attractive to the frustrated decision maker, the proposition that "if it is simple, it must be correct" is a clear error in logic. The appeal is to simplicity and the easily understood, not to the issues themselves.

The neutralization of the modern integrated air defense system by SEAD weapons/tactics and the criteria that assess the neutralization's effectiveness promise to be increasingly complex. The range of complexity is increasing while the commonsense approach's impetus attempts to force simplicity. "Neutralizing" a SEAD target in the future will cover a large range of options. Destruction and brute-force jamming are no longer the only means available. More sophisticated air defense systems employing multiple sensor arrays and tracking systems netted together by an advanced information management mechanism, because of their complexity, offer many more vulnerabilities to exploit for the SEAD tactician. The combinations of weapons and tactics that might possibly "work" against the modern IADS can be quite complex. The commonsense approach can militate against a realistic appraisal of alternatives because they are "too complicated," leading decision makers to opt instead for the simple, straightforward answer. This limits the ranges of more relevant means of assessing SEAD effectiveness by restricting criteria development to superficial goals and objectives.

Commonsense approaches in SEAD criteria development have usually been the results of two separate kinds of behavior, both related to decision makers. The first is a response to being surprised by events that do not fit the commonsense paradigm in vogue. The second response is the result of decision-maker frustration when directed action fails to yield expected outcomes.

The SAM defense nets of North Vietnam and the integrated air defense structures the North Vietnamese developed to counter American air power, while known quantities, were technical surprises to decision makers. The information concerning the North Vietnamese capability that decision makers dealt with did not fit their view of the world at the time. Only the reality of mounting American aircraft losses to the enemy IADS net forced a response. The word went out to "solve the problem" without decision makers knowing precisely what the problem was. The result was a string of separate, reactionary responses that solved the immediate, short-term problem at the expense of assessing the evolutionary development of a more and more complex enemy air defense system.

The second behavioral response that results in the appeal to the commonsense approach is generated by the decision maker's frustration when SEAD options do not meet expectations. The SEAD weapons programs of the late 1980s suffered badly as a result of this response. The upgrade programs for the F-4G and EF-111, the improved internally mounted self-protection jammers for the F-111 and F-15, the program to acquire a follow-on Wild Weasel replacement for the F-4G, and the advanced follow-on integrated electronic warfare suites of advanced bombers and fighters were severely curtailed—or even cancelled in some cases—because of this frustration. The frustration—and the apparent compulsion to find a simpler, commonsense approach—was the result of poor communication, unrealistic expectations, and obsolete paradigms as much as it was the result of complexity. Development of sound SEAD criteria in an atmosphere charged with such frustration led to a destructive spiral between decision makers and the engineering community. The commonsense criteria developed from frustration created measurements of SEAD effectiveness that did not adequately address the complexity of SEAD weapons and tactics or the way SEAD fits into overall modern military strategy and tactics. This yielded SEAD approaches and weapons choices that also fell short of expectations, leading yet again to another round of "commonsense" solutions.

Communication between the electronic combat technician who has been charged with developing SEAD measures of effectiveness and the decision maker charged with criteria formation has been hampered by the inherent complexity of the subject. Technicians' single-minded focus on very specific elements and the failure of the decision makers throughout the bureaucratic structure to streamline communication channels have resulted in significant disconnects. The case of the EF-111 serves as an example. The specifications for the EF-111 improvement program were related to technical system performance, not overall capability. The failure to deliver a high-risk, beyond state-of-the-art piece of equipment for the EF-111 upgrade on time (it was actually three and a half years late) was the root reason the Air Force cancelled the contract.[11] In retrospect, however, an alternative engineering architecture using state-of-the-art computers could have possibly achieved operational and objective goals for the EF-111 upgrade, but the decision makers' frustration at complex technicalities and the failure of the engineering community to deliver what it had promised resulted in a backlash of "return to basic commonsense" approaches.

There was a disconnect between the user community, which established SEAD requirements for the EF-111, and the engineering community to understand what the user "really" needed. Decision makers correctly interpreted the disconnect as a failure of the engineering approach to provide overall criteria for SEAD effectiveness but incorrectly determined its cause as unnecessary complexity and technicality. The result was the backlash commonsense approach. Though not totally responsible for the cancellation of some very critical SEAD programs that may have far-reaching impact on US SEAD capability in the future, the return to "old-fashioned common sense" may have done far more harm than good in the long term.

Each of these hindrances to communication—as with the EF-111 and F-4G upgrade programs—has resulted in decision makers becoming frustrated and resorting to a system that "makes sense" in the short term. The damaging element in the commonsense approach is that it tends to inhibit a broader view and excludes more sophisticated treatment of the criteria.

As SEAD systems and tactics become more encompassing and technical, they become increasingly complex. The complexity may have reached a point where decision makers feel defeated before they start. When this occurs, decision makers often resort to the commonsense approach out of desperation and the half-whimsical belief that "if it's that complex for us, the enemy will never understand it either."

The commonsense approach does serve a very useful function. It tends to focus the technician's attention on the task as well as clarify the objectives of the decision maker. Unfortunately, what makes sense to the decision maker may not make sense to the tactician or technician. While the commonsense approach may lead to the development of criteria that is counter to long-range interests in the pursuit of short-range goals, it can also serve as a catalyst for needed dialogue between the tasker and the tasked.

In short, the commonsense model is both a curse and a blessing. The appeal to common sense serves to focus a community on basic objectives and to build a consensus for future directions. However, as a *standard* of measure for SEAD, it is especially insidious because requiring it curtails more sophisticated treatment of the SEAD problem. The commonsense approach continues to be a convenient tool in applying pragmatism, but it is dangerous as a primary standard of measurement.

The Objective-based Approach

The objective-based approach not only embodies elements of each of the previous approaches as tools but also suggests that any criterion or measurement of effectiveness be related to overall objectives that are traceable throughout the entire process. The objective-based approach is a commonsense test in that it answers fundamental questions about SEAD's purpose and role in the application of military power in clear, easily understood ways. The objective-based approach is empirical in that many of SEAD's tools have measurable effects and is historical because it compares a particular tactic, weapon, or strategy that has been used before to determine whether it is applicable—and effective—for potential SEAD use. Most importantly—because it is all encompassing—the objective-based approach provides a more adequate contextual framework with which to assess *joint* SEAD in the joint war-fighting arena.

The objective-based approach is—like the historical, the engineering, and the commonsense approaches—a product of an evolutionary process. It is not new; rather it is the most current manifestation of a continuing process which seeks to achieve equilibrium with its surroundings. SEAD has undergone dramatic changes in a world which has also undergone dramatic changes. Finding a SEAD criterion that relates to the real world and measuring SEAD effectiveness in relevant terms in an altered environment call basic objectives into question. The objective-based approach focuses the process on basic questions. What is it all about? What is it that needs to be done? What is the best way to do it? Does everyone know what to do?

The objective-based approach forces those who provide objective guidance into a continuous cycle of reclarifying and defining precisely what they want done; it forces those who execute the strategies into a continuous cycle of ensuring that their equipment and tactics meet the objectives. *The standard of measure is denominated in terms of the objective.* The criteria developed using the objective-based approach "make sense" throughout the decision-making process from national-level decision makers charged with defining national security policy to war fighters using the SEAD weapons and tactics developed from the process.

Figure 6 places the objective-based approach in relation to the three previous approaches. The evolutionary blends are discernible over time. The graph shows that the SEAD criteria development was initially propelled by historical comparison. There were elements of each of the other approaches present with the historical approach. As the process evolved, historical

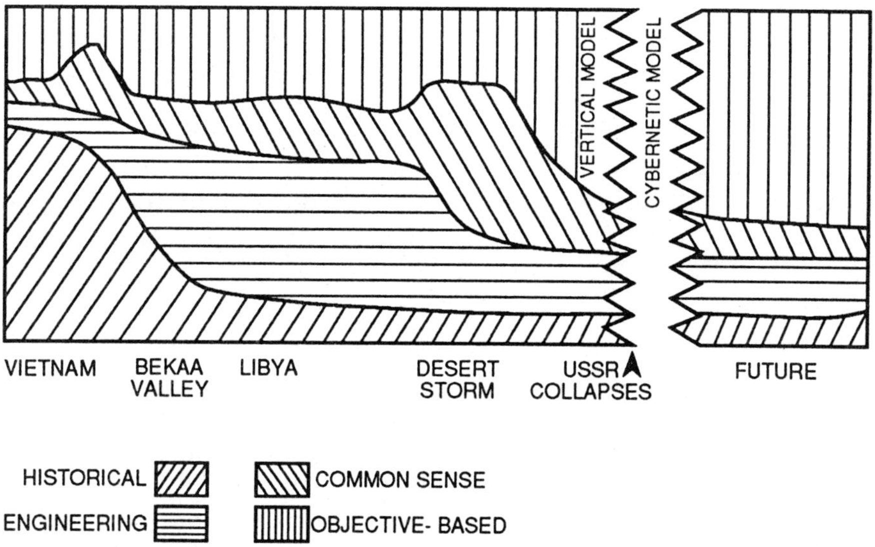

Figure 6. Evolution of criteria/MOE approaches.

approaches were gradually replaced with engineering approaches. Technology propelled the process so quickly that it rapidly outstripped decision makers' ability to control it. This eventually resulted in a backlash reversion to commonsense approaches.

The commonsense approach used to develop Vietnam SEAD criteria, however, differs qualitatively from that used in the late 1980s. The commonsense approach used in Vietnam was a function of the historical approach, while the one used in the late 1980s was a result of decision-maker frustration. In the case of Vietnam, the commonsense approach led to a greater use of empirical methodology. In the backlash case of the late 1980s, the commonsense approach forced the SEAD process of criteria development and measurement of effectiveness to evolve into the objective-based approach.

Table 6 outlines the characteristics of each of the approaches. Many of these features reflect the evolutionary trends outlined in chapter 1. The objective-based approach, however, differs in major respects from all the others: it is proactive and prescriptive. Because it is process oriented, the objective-based approach focuses on how the decision-making process aids in the development of SEAD criteria and measures of effectiveness. Because it is all encompassing, the objective-based approach focuses on JSEAD applications as opposed to single-service SEAD employment options. Because it is decision maker-driven, the objective-based approach stresses the requirement for clear channels of communication and a streamlined chain of command with clear lines of authority.

Table 6

Comparison of Cold War to Post-Cold War SEAD Environment

Cold War	Transition	New Environment
– Clear threat – Resource-rich – – Specific weapons for specific targets – – Service oriented – Technology dependent – Deterrent defensive strategies – Relative economic health – Political support	– Threat perceived as gone; undergoing changes – Resources dwindling – – Residual assets available during Desert Storm – – Integration of services – Technology gap closing – Collective coalition security strategies desired – Economy shaky – Political support ambiguous	– Threat exists at same or higher level – – Diffuse – – Sophisticated – Resources scarce – – Available assets must perform multiple roles – – Joint warfighting a requirement – US overwhelming force application unlikely – Technology gap close – Collective/coalition strategies required – Economy integrated with world – – Interdependent – – Unstable – – Less able to control – Political support ambiguous

The major impetus for the objective-based approach was the demise of the Soviet Union as a threat. Even more than Desert Storm, the sudden removal of the Soviet threat as a key element of national security policy forced a radical reappraisal of how we measured military worth—especially as far as defense suppression was concerned. The sudden demise of the Soviet threat was *the* watershed event which forced a qualitative change in the emerging objective-based approach to SEAD criteria/MOE development. The reappraisal had far-reaching impacts on SEAD tactics and weapons development as a function of war-fighting strategy. Table 7 compares cold war SEAD criteria/MOE practices to the uncertain world of the future.

Until now, the use of the vertical, linear model defining the SEAD criteria/MOE development process—which begins with

Table 7

Comparison of Critical MOE Approaches

Historical	Engineering	Common-Sense	Objective-based
– Used when new situation encountered	– Empirical	– Assures universal frame of reference exists	– All encompassing
– Reactionary	– Specific	– Reactionary	– Relates means to goals
– Solves specific problem	– Influenced by technology	– Focuses on specifics	– Proactive
– Short-sighted	– Task-oriented	– Goal-oriented	– Deductive
– Influenced by doctrine	– Not involved in decision-making process	– Seeks consensus	– Stresses continuity of information flow
– Less sensitive to technological change	– Neither reactionary nor proactive	—Seeks to simplify	– Process-orientated
– Not predictive	– Piecemeal application	– Integrative	– Objective-driven
– Deductive	– Threat-based	– Threat-based	– Seeks best overall solution
– Piecemeal application		– Deductive	– Integrative
– Threat-based		– Short-sighted	– Capability-based
			– Long-term goals
			– Influenced by overall goals

leadership defining objectives and ends with the execution phase, which achieves those objectives through successive stages—has served the electronic combat community well. However, the integrative forces of SEAD evolution and the speed of the technology/communications revolution now bring the relevance of the vertical, linear model of the process into question. Figure 7 compares the traditional vertical model to "reality." The vertical, linear depiction of the criteria/MOE development process is a product of an earlier time, and it responds to a paradigm that no longer fits the facts. The vertical model reflects a paradigm that is superfluous to the emerging new world. Attempts to modify or to restructure it but retain its linear, iterative nature hinder the evolution of the SEAD criteria/MOE development process.

What model, one might ask, does "fit reality" in the attempt to build the proactive, objective-based criteria model needed for determining future SEAD options? Needed is a process that

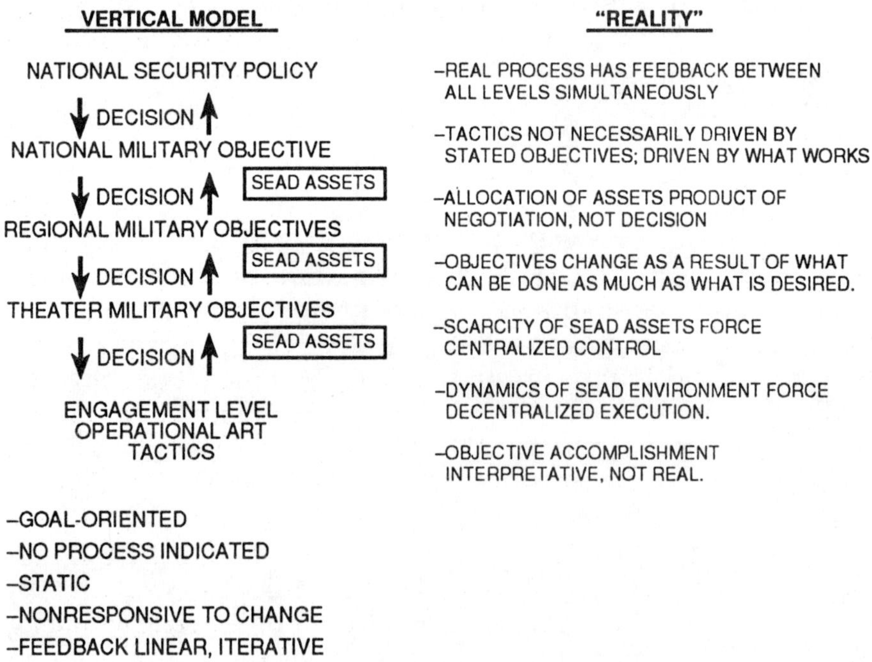

Figure 7. Traditional vertical model.

focuses on dynamics, accommodates change, transfers information fluidly from one portion of the process to the next *regardless* of direction, and relates relevance of activity to objective accomplishment.

The objective-based SEAD criteria of the future will be process-oriented as well as goal-oriented. This is required because the potential threat facing future US SEAD assets is growing in sophistication, size, and composition. The range of capabilities represented by the new threat will require a SEAD criteria/MOE process that can assess the threat's capabilities and select the most appropriate aggregate weapons and tactics mix to meet it—and the process must respond rapidly to be meaningful. The process will need to be self-adjusting to unanticipated changes and events.

Of the many models available in the physical, engineering, and social sciences, the one which comes closest to combining what the objective-based SEAD criteria/MOE model needs to be with the evolving reality of the future world is Karl Deutsch's cybernetic model. The model was introduced in the late 1960s to explain the function of feedback in the political decision-making process. The model was an attempt to depict the processes as they *really* occurred, not as decision makers idealized them. The cybernetic model is well suited for a SEAD objective-based model for many reasons. It is process-oriented, it characterizes objectives as the primary drivers of the process, and it describes a self-adjusting process where equilibrium is achieved when all of the individual parts of the process have contributed to achieve the overall objective. Figure 8 compares the JSEAD objective-based cybernetic process to the reality around it. One can also compare the vertical model in figure 7 to the cybernetic model to illustrate the apparent shift from a reactionary cold-war paradigm to the more proactive, process-oriented, objective-driven paradigm.

The cybernetic process is primarily a political science construct, and applying it to the technical world of JSEAD and joint war fighting is difficult. But its relevance to SEAD's situation is compelling. Several situations currently facing JSEAD decision makers serve as examples of why we need an objective-based approach. First, the basic mental construct of viewing reality (paradigm) used to address JSEAD concerns

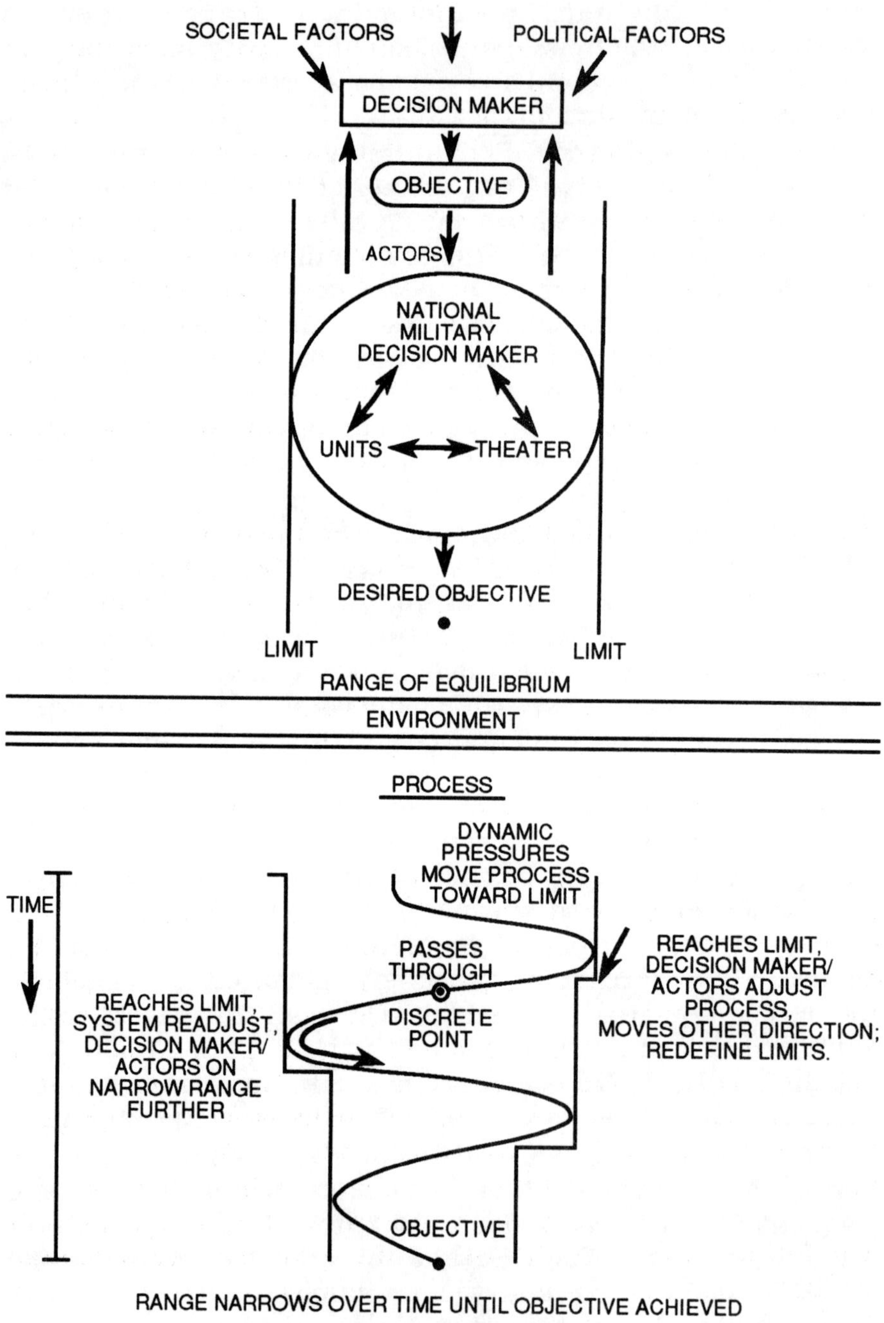

Figure 8. Cybernetic model.

lags behind reality. More disturbing, there is positive resistance to adopting a more adequate paradigm. Second, the requirement statements for technically advanced SEAD systems to meet the proliferating and expanding technical threat of the future are couched in cold-war concepts and jargon. Third, the requirement for true joint approaches to the SEAD problem are bogged down in service parochialisms and disputes over authority. Fourth, decision makers have not addressed criteria which meet *joint* SEAD requirements.

The SEAD criteria currently used are defined in archaic terms that have little to do with the reality of the uncertain world. Current SEAD criteria reflect service-oriented SEAD concepts that militate against measurements of effectiveness that would test combined uses of SEAD weapons and tactics to achieve overall objectives.

In Deutsch's modified cybernetic model, the overall objective is the primary input driving the process. The process achieves equilibrium when the objective is being met. The objective-based approach to developing JSEAD criteria recognizes decision making as a process where all the parts *work together* to achieve overall equilibrium. JSEAD objectives are driven by an accurate appraisal of reality, an appreciation of the world as it really is, not as a decision-maker's decree. Therefore, the first paradigm that must be discarded is the vertical, linear model of decision making.

The revolution in communications technology and information transfer has made data available to the entire decision-making structure simultaneously. This makes the opportunity for feedback a constant reality. As a consequence, the decision-maker's responsibility in the JSEAD process has shifted from one of directing to one of determining whether or not the objectives are being met. The criteria used to evaluate whether the objectives are being met is driven by the objectives themselves, not by the process. Agencies within the process adjust their activities with one another to achieve the objective using a criterion that defines measures of effectiveness that are quantitative, qualitative, and dynamic. This means establishing priorities and the conditions under which they might change. By ensuring feedback throughout the process, decision makers and process participants can maintain dynamic equilibrium and adjust to changing circumstances. Given criteria that specify objectives in unambiguous language, decision makers and process participants can develop measurements of effectiveness that define whether or not objectives are being met and then can alter their activities accordingly.

Suppression of enemy air defenses during Desert Storm illustrates the transition to the cybernetic paradigm. At the macro level, Desert Storm SEAD objectives were encompassing and broad ranging. In fact, the SEAD objectives were so broad ranging that a SEAD campaign plan had to be drawn up. Strategists did not consider Iraqi SEAD targets as individual nodes or specific locations linked in a sequential, linear manner to a master command-and-control center. All SEAD targets were considered in light of their contribution to the whole Iraqi IADS—as part of a system. In other words, the objective was the degradation of the entire system. Rather than attack the Baghdad SAM complex first, which a linear "high-value target" thought process would lead a Vietnam-era tactician to do, the first SEAD target in Iraq was a radar complex located on the southern Iraqi border. Its destruction denied the Iraqi C^3 structure information. The objective was the degradation of the total system, not the destruction of high-value SAM sites. Objective-based criteria in the Iraqi campaign, however, were spotty. In the translation from macro concept to tactical application, several disconnects were noticeable.

First, tactical fighters discovered the need for *tactical* suppression of enemy battlefield air defenses. This was especially true for A-10s, F/A-18s, and F-16s flying in the daytime in northern Kuwait and the Basra region of southern Iraq. Oddly, tactical suppression of organic air defense—other than hostile, emitting radar trackers which were quickly quieted with HARM-shooting Air Force, Marine, and Navy aircraft—was not considered in the original SEAD campaign. Iraqi gunners and SAM operators—even though they had successfully been severed from the Iraqi command and control net—continued to operate autonomously and often blindly. The fact that Iraqi fire was grossly inaccurate was of no great comfort to the average coalition aircrew that observed multiple

SAM launches and AAA bursts in their general direction. Hence, members of fighter missions, amongst themselves, designated specific sorties within their flights to take on a tactical defense suppression role. Using the strict vertical, linear model, the development of such a tactic cannot be explained easily. Where did the direction come from? How did the decision-making process work that resulted in the ad hoc creation of the tactic? Using the cybernetic model easily explains the development of the ad hoc tactical defense suppression tactic. The objective-driven process needed to make an adjustment to reach equilibrium, and the development of the ad hoc tactic was it.

The second disconnect was the failure to recognize paralysis of the enemy as a valid SEAD objective. The SEAD campaign was enormously successful, but coalition tacticians, in identifying target objectives with great precision, did not identify a precise means of determining how successful their tactics and weapons were. Consequently, there was no way to tell how effective SEAD weapons and tactics were because analysts were not sure what they were looking for. It was not until after Desert Storm that analysts discovered that the primary effect of coalition SEAD was paralysis of the Iraqi IADS and C^3 system. Paralyzing a system that depends almost entirely on command and control for effective use of the IADS, as many analysts pointed out afterwards, is virtually as effective as destroying the collective IADS weapons themselves—provided that the paralysis is maintained over the course of the war. The initial massive assaults on selected critical IADS and C^3 nodes shocked the Iraqi IADS into paralysis, and persistent attacks on the IADS structure 24 hours a day ensured the Iraqi system never recovered sufficiently to be effective. The important point to remember here is that analysts identified paralysis of the Iraqi IADS as a valid SEAD objective *after the fact.*[12]

The evolution to objective-based criteria using a cybernetic paradigm will overcome disconnects like the ones mentioned. More importantly, objective-based criteria provide a basis for proactive weapons and tactics development as well as a framework with which to view the overall SEAD process.

Setting the Context: Continuums as Assessment Tools

SEAD effectiveness criteria historically have been specific, empirical, and related to tactics. The realities of the emerging world order and the uncertainties it brings make this methodology increasingly suspect. We may have reached the point where the traditional methods of analysis are applied to conditions that no longer pertain.

The 10 variables traced in chapter one can be distilled into four evolutionary continuums. These continuums are

1. piecemeal/integrated,
2. need-based/resource-based,
3. threat-based/capability-based, and
4. defensive/offensive.

The continuums give us a sense of historical perspective in SEAD development, a way of assessing where we are going and what forces are at work in shaping future SEAD concepts. They represent the environment in which the cybernetic process operates. Figure 9 shows the cybernetic process superimposed on the continuums scale as SEAD objectives—and the process itself—attempt to reach equilibrium in rapidly changing realities. The evolutionary trend forces the process to change its objectives based on reality. Not only are the continuums forcing the process to adapt rapidly, but the continuums themselves change at different rates.

The four continuums attempt to assess the changing context of criteria development. Clearly, if the old paradigms no longer apply, their continued use will yield—at the very best—indeterminate measures of overall effectiveness. The continuums attempt to place specific types of measurement and criteria selection in a more relevant construct.

The piecemeal/integrated continuum relates to the way JSEAD assets are employed in combat. Sometimes resource limitations, doctrinal predisposition, or situation-dependent factors might dictate the use of forces along lines indicated by this continuum. For instance, defensive applications of some JSEAD assets have usually been examples of the piecemeal

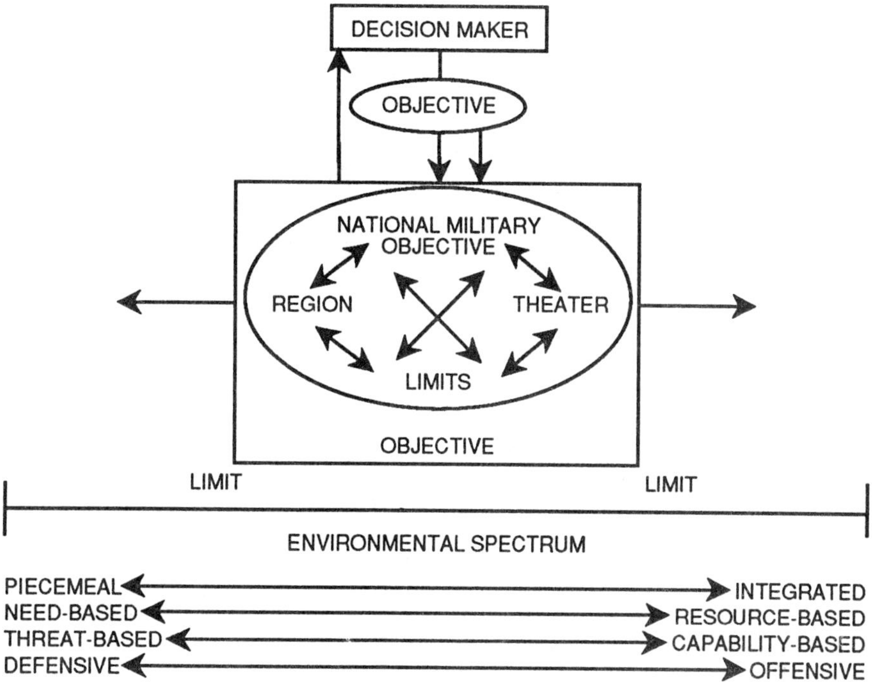

Figure 9. The cybernetic process and the four continuums.

employment approach. Using this approach, F-4Gs would be tasked a few at a time to defend specific aircraft elements, or Navy EA-6s would be tasked for specific defense of a group of aircraft entering a target area. Piecemeal employment of the F-4G, F/A-18 HARM-shooter, EA-6B, and the EF-111 were standard practice—when used in a defensive mode—during Desert Storm. When resources are available, the piecemeal approach is an efficient way of gaining maximum defense for high-value assets.

When resources are limited, using the piecemeal approach forces the battlefield commander to prioritize missions by sorting out who is to be defended and who is not. Alternatively, the commander may be forced to select a different way of employing JSEAD assets to make the most of their abilities.

Offensive applications of JSEAD tend to present an integrated approach to employing the assets. The collective and integrated efforts of many different types of weapons systems to neutralize an enemy war-fighting structure, for example, lead to a more integrated approach in achieving the objective.

Achieving the SEAD objective in rapidly changing circumstances sets up a relationship among desired SEAD outcomes, feedback loops, and the environment. The piecemeal approach is gradually giving way to a more integrated method of SEAD application because of fiscal constraint, limited resources, technological sophistication, immediate access to pertinent information, and societal pressures.

The key point is that SEAD's cybernetic process will seek an equilibrium to the evolutionary pressures pushing the continuum toward the integrated end of the spectrum. Just because the equilibrium reached may not be the equilibrium desired does not mean that the process is dysfunctional; in the absence of proactive direction, the process will make whatever adjustment to reality fits the situation.

The integrated approach represents the evolutionary trend. Continued piecemeal applications, no matter how desirable they may be, must be viewed in the reality of the evolutionary trend. Figure 10 shows the evolutionary trend from piecemeal to integrated and its relative impact on the SEAD criteria-

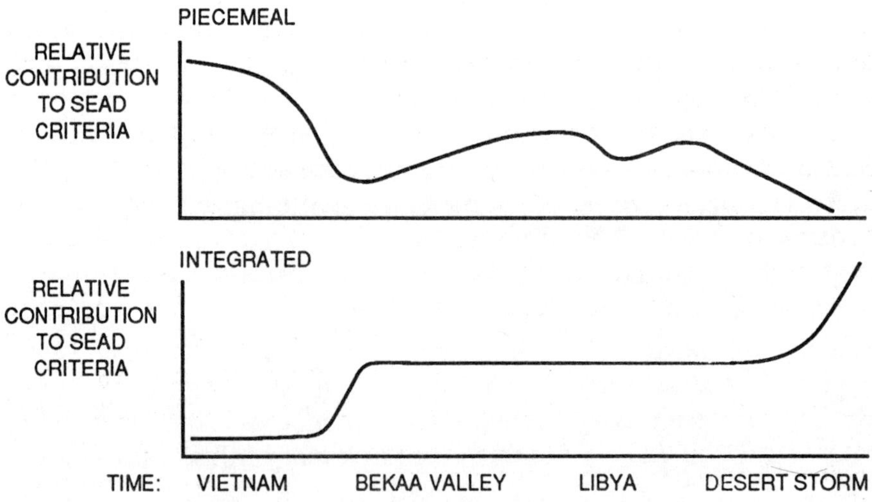

Figure 10. Piecemeal/integrated continuum.

development process. Objectives need to be defined as a function of the integrated use of concepts and SEAD weapons not only because doing this makes sense but because it represents the most appropriate adaptation to changing realities.

The need-based/resource-based continuum represents two different approaches to resources and objectives that can be graphed, roughly, as a spectrum. On one side is the question: "Given the objective, what resources do I *need* to accomplish the objective?" On the other side is the question, "Given the *resources* I have—and am likely to receive—what can I do to achieve the stated objective?"

The continuum's effect on JSEAD becomes more relevant in a resource-constrained environment. JSEAD has been moving steadily toward the resource-based end of the continuum. Very specific requirements were driving the acquisition of specific systems to fill the need. In a resource-rich environment, filling specific needs made more sense. A justifiable military need drove the system to acquire required resources. From these, the users derived very distinctive tactics and weapon combinations to deal with the threat.

An altered political and economic environment forced a change in the continuum. The military need no longer carries the priority it once did, and weapons and tactics employment concepts are becoming objective-driven. SEAD is especially hard hit by this change. The general military threat is diminished, but the specific SEAD threat of the future continues to grow.

Evolutionary trends show that, no matter what the threat, SEAD objectives must be tempered with a recognition of the facts. Even as SEAD resources become more scarce, the requirement to perform the SEAD mission in a more sophisticated, threatening electromagnetic environment remains. This paradox creates a significant challenge for SEAD criteria development. Rather than make the blanket statement that US forces need to maintain the technological edge, the SEAD community needs to specify *what kind of edge* is needed given the resources available. Rather than ask for general technological improvement programs, the SEAD community needs to assess how it can use what resources it is likely to receive to achieve objectives. More pertinently, the objectives themselves may need to be redefined in terms of what is achievable with the given resources. The fact is that future SEAD may not be able to achieve objectives as they are currently understood. The resource-constrained environment will force integration of SEAD resources and shift the emphasis from specific threat coverage to general capabilities against a broad spectrum of air defense threats.

Figure 11 shows the need-based/resource-based continuum and its impact on objective-based criteria development. The shift from criteria developed in need-driven, resource-rich surroundings for force structure and weapons acquisition to a resource-constrained environment that is forced to use what it has to achieve SEAD objectives is noteworthy. Desert Storm, as dramatic an event as it was, was not the watershed event for this continuum. The critical event was the demise of the Soviet Union, and its impact has accelerated the evolution of this continuum toward the resource-based end of the spectrum. Clear objectives are impossible to define because the transition from a need-based to a resource-based SEAD environment was virtually instantaneous. The shock to the process used in determining SEAD effectiveness is numbing; the system is out of balance and will remain so until decision makers and the SEAD community can define the relationship

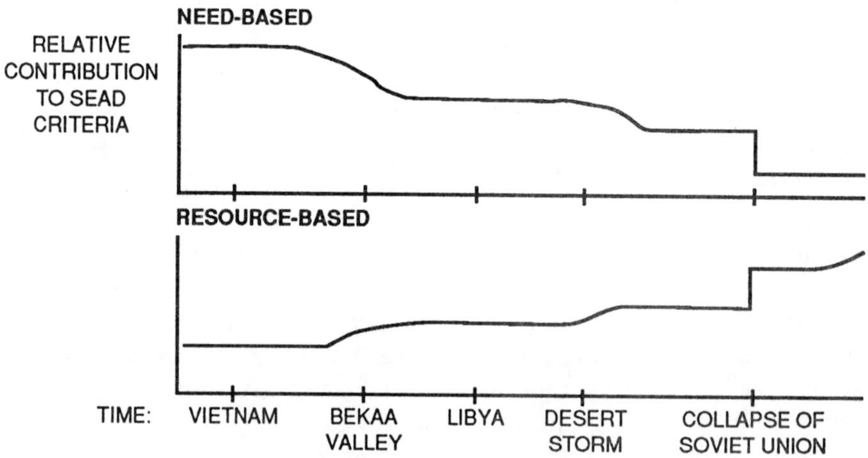

Figure 11. Need-based/resource-based continuum.

between achievable SEAD objectives and adequate SEAD resources in concrete terms.

The threat-based/capability-based continuum is closely related to resource availability and the changing nature of the air defense threat. As it attempts to define achievable objectives and a means of determining effectiveness in a resource-constrained environment, the SEAD community has been forced to discontinue developing tactics and weapons based on specific, identifiable enemy systems and to begin using a method which acknowledges the proliferation of sophisticated air defense technology among many potential enemies. Rather than Soviet-style weapons and the massive Central European land-war scenarios of the cold war with troops and weapons in place, the US SEAD community faces an unknown potential enemy that can field the best weapons Western arms industries are capable of selling them. Rather than in-place forces, the US SEAD community must be able to deploy a defense suppression capability as soon as requested. Air power, as a critical determinant of national power projection, cannot be employed prudently without the protection offered by SEAD.

The movement from threat-based to capability-based tactics and weapons is a general evolutionary trend faced by the entire military, but it is especially problematic for SEAD. EF-111/EA-6B jamming works against specific radars. Antiradiation missiles have capabilities against a specific range of emitting targets. Bombers and fighters have differing—and specific—susceptibilities to enemy air defense nets. Intelligence collection, target selection, and damage assessment for SEAD objectives are specific. There is very little that is "general" about achieving SEAD objectives. Yet the evolutionary trend propels the SEAD community toward acquiring a capability-based SEAD force. How the SEAD community will develop capability-based objectives and a means of measuring effectiveness promises to be vexatious.

Figure 12 shows the impact the evolutionary trend toward capability-based SEAD weapons and tactics has on the development of objective-based criteria. The major impact relates to threat definition. Defining the threat in terms specific enough for tacticians to know what weapons to use in what circumstances while retaining SEAD force structure and

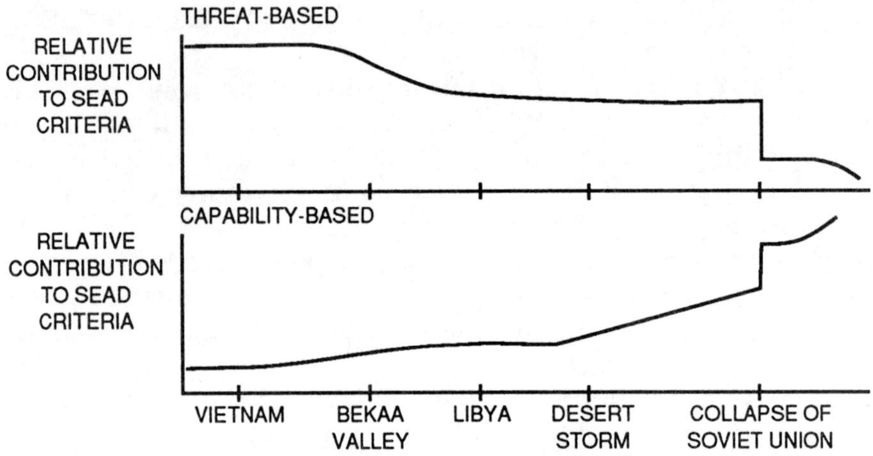

Figure 12. Threat-based/capability-based continuum.

strategies that encompass a broad range of capabilities will be a challenge.

The depiction of the threat-based/capability-based continuum is the same as that of the need-based/resource-based continuum in general form. The other two continuums are reasonably linear and smooth in their transitions from one side to the other. Threat-based forces, strategies, and tactics versus their capability-based counterparts have a more distinctive nature: they tend to be one or the other. Capability-based and resource-based SEAD approaches are directly related. The combination of a perceived diminution of the threat and subsequent loss of resources has accentuated trends toward the capability-based and resource-based end of the spectrum—regardless of the true nature of the threat SEAD forces face. The attempt to make the transition smooth and linear may be more indicative of hopeful outcomes than of concrete reality. Nonetheless, the abstraction is useful as a tool for developing JSEAD criteria and assessing effectiveness.

The defensive/offensive continuum reflects the movement of SEAD from its original orientation in a defensive support role for protecting friendly aircraft to its current status as both a defender of air power and an offensive weapon in its own right. The movement has been inexorably pushed toward the offensive end of the spectrum by the logic of technology and superior firepower. Advanced antiradiation missile design, advances in electromagnetic power generation, remotely piloted vehicles, unmanned aerospace vehicles, stealth fighters and bombers, enhanced communications technology, true 24-hour-a-day attack capability, precision-guided munitions, and the growth of space-based surveillance and targeting systems buttressed by digital computer technology have created SEAD capabilities undreamed of even 10 years ago. These capabilities have far more implications for enhancing power projection than they do for solidifying defensive positions. Hence, there has been a positive movement toward the development of offensive uses of SEAD that has, until recently, been unchecked.

Assessing the defensive/offensive continuum's impact on the development of objective-based criteria for measuring SEAD effectiveness is difficult. The terms *defensive* and *offensive* represent war-fighting concepts as well as describe a tactical situation faced by one side or the other in combat. While it is clear that technology has given SEAD a distinct boost toward the offensive end of the spectrum, changes in the politico-military environment seem to have slowed the pace. Resource constraints and the ambiguous nature of the future air defense threat, coupled with pressures to integrate SEAD plans into a single, cohesive plan, have slowed the evolution of SEAD's offensive dimension.

The objective-based criteria for SEAD may call for either a defensive or offensive application depending on the situation US forces face. The current military proclivity indicates a strong preference for decisive use of military power to overwhelm an enemy. Desert Storm SEAD application was a premier example of using overwhelming technology and superior firepower to neutralize an air defense system. It was also an apt example of the use of SEAD as an evolved offensive concept. Yet US reluctance to provide an adequate tactical defense suppression over the battlefield shows that strategic SEAD planners may have gone too far toward the offensive side of the spectrum. Fortunately, there were few coalition aircraft losses to Iraqi battlefield air defense systems. Nonetheless, the judicious use of the defensive/offensive spectrum might provide a way of bringing SEAD objectives into clearer focus and provide a broader perspective against which to measure SEAD effects.

Figure 13 shows the defensive/offensive continuum—a variation of the Clausewitzian model.[13] The primary difference is that Clausewitz used four discrete categories rather than a continuum stretching across a spectrum. SEAD's evolution from defensive to offensive has been very rapid until recently. The movement in the piecemeal/integrated SEAD continuum toward integrative uses of SEAD has enabled more offensive applications of SEAD weapons and concepts. The movement to the integrative end of the spectrum provided the impetus to bring the separate SEAD weapons and concepts together into a coherent whole, and, for the first time, tacticians could see the possibilities SEAD provided as a total concept for the offensive. The movement toward integrative methods also began the incipient movement towards the development of joint SEAD objectives and criteria for measuring effectiveness.

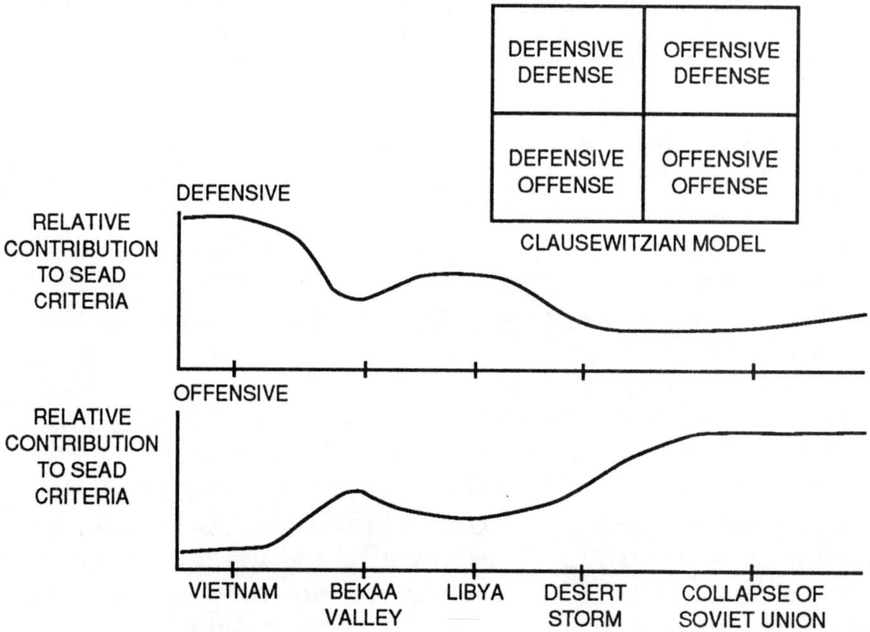

Figure 13. Defensive/offensive continuum and Clausewitzian model.

At the same time, the transition to a capability-based and resource-based orientation for SEAD have slowed the development of offensive SEAD concepts because of a slowdown in technology research and in weapons procurement. The net result of these forces is a gradual slowing of SEAD's evolution toward the offensive end of the spectrum.

Challenges for Objective-based Criteria

The four continuums in chapter 1 serve as a way of converting the 10 variables into evolutionary trends. They provide a sliding scale for viewing objective-based criteria development. Nonetheless, the objective-based criteria must undergo the same "reality checks" that every emerging scheme experiences. The challenges are daunting.

The relationship between technology and SEAD development has been crucial. The challenge for objective-based criteria will be in the formation of proactive criteria—criteria that not only can select the general areas of research required for maintaining a US lead in critical electronic combat technology but that can define *the kind of edge* American SEAD will need to have in the future.

The gradual withering away of cold-war force structures will have enormous impact on the general ability of the military to project American power when and where it is needed. Air power's stock as a key component of national military power will grow in the new world order—a world order which will value prompt, decisive responses to crisis situations. The US is not the only nation to recognize the importance of air power. The lessons of Desert Storm taught observers as well as participants, and SEAD's lesson was clear. With SEAD, modern air power can cripple a nation's ability to wage war. Without it, air power's effectiveness is greatly diminished.

The challenge is to determine how much SEAD force structure is enough to meet the challenge posed by the growing air defense capability of the potential enemy. How much SEAD force structure represents "overwhelming" capability? What mix of SEAD weapons will be adequate to protect American air power and cripple the enemy's ability to defend from air attack?

Continued fiscal constraint and the accompanying military drawdown have already forced a major reorganization among the services. Many more changes will follow. The challenge in developing objective-based SEAD criteria will be to create or adapt organizations to accommodate the reality of "doing more with less." Rather than wait for the decision-making process to decree the shape of future SEAD organization, SEAD tacticians and practitioners must take an active part in the creation of the reorganized American military to ensure SEAD objectives can be met.

Military doctrine—both single-service doctrine and joint doctrine—poses a huge challenge to SEAD. As SEAD criteria development shifts more and more to an objective-based approach, SEAD tacticians and strategists have an excellent opportunity to take the lead in determining how SEAD can best be used to achieve overall objectives.

Military doctrine is undergoing significant change. SEAD practitioners need to get in "on the ground floor" to establish criteria for developing effective defense suppression measures that relate directly to overall objectives rather than to the narrow ranges dictated by formulators of doctrine, who have little understanding of SEAD's contribution to the war-fighting effort.

The tactics of Desert Storm—while vindicating much of the investment of SEAD technology—will gradually fade as a model because the weapons, the expertise, and the organizational structure supporting them will change radically. Objective-based criteria will help the changed SEAD force use its capabilities to best advantage by balancing what *can* be done with what *needs* to be done in accomplishing its tasks. Pragmatism and innovation will be key to the process.

The greatest challenge for emerging objective-based criteria will be to relate overall strategy to concrete activity. The setting of clear SEAD objectives and the constant refining that will take place must take into account far more variables than the current paradigm allows for. Improvements in managing SEAD intelligence information and its timely distribution will be required for the emerging objective-based criteria to take hold and make a meaningful difference.

The hallmark of the emerging new age seems to be the growth of information—its management and distribution—as a determinant of power. As a direct consequence of the age of computers, information itself may become the key to power in the twenty-first century. The nation that can control data, deny it to others, or willfully shape it will be preeminent in the wars of the future. In this context, SEAD criteria may be couched more in terms of information denial and shaping than in the more traditional concepts of radar destruction and jamming. Whatever the outcome, in order for SEAD criteria or measures of effectiveness to be meaningful in the uncertain world of the future, they must show clear linkage between the objective and the desired result.

Notes

1. Karl Deutsch, *The Nerves of Government* (New York: The Free Press, 1966), 75–94. The cybernetic model Deutsch presents has undergone considerable change since its original introduction, but the concept of a process which seeks equilibrium through self-adjustment is an extremely useful one in criteria development.
2. Karl R. Popper, *The Poverty of Historicism* (London: Routledge & Kegan Paul, 1976), v.
3. Ibid.
4. US Air Force Operational Test and Evaluation Center (AFOTEC), *Proposal for Determination and Application of Measures of Effectiveness* (Kirtland AFB, New Mex.: AFOTEC, 1991), 3.
5. Ibid., 4.
6. Ibid.
7. AFOTEC, "OAY/OAN Point Paper on TAWC [Tactical Air Warfare Center] Response to AFEWC [Air Force Electronic Warfare Center] ASPJ [Advanced Self-Protection Jammer] MOE Study," (Kirtland AFB, New Mex.: AFOTEC, 1991), 2.
8. AFOTEC, *Proposal for Determination and Application of Measures of Effectiveness*, 4.
9. AFOTEC, "Air Force Cost and Operational Effectiveness (COEA) Process," briefing to AFOTEC/OA, February 1992, slide no. 7-2.
10. William Graham Sumner, *Folkways* (Boston, Mass.: Ginn and Company, 1906), 521–30. Sumner introduces the concept of "mores" in his book, which lays the foundation for the concept of "common sense" and the potential dangers it implies.
11. The author was the EF-111 functional manager for the Tactical Air Forces (Tactical Air Command, United States Air Forces Europe, Pacific Air

Forces) during this time and was deeply involved with attempts to upgrade the EF-111.

12. This is an example of the uses of hindsight to discover the "objective behind the objective." It may be an unfair criticism of the SEAD planning for Desert Storm because no one at that time could have predicted the synergistic effects of simultaneous attacks on the Iraqi IADS and the effect of subsequent around-the-clock harassment of the IADS/C^3 structure to paralyze it using the SEAD criteria of the time. The challenge for objective-based criteria is to foresee, as much as possible, the unforeseen objectives as a way of economizing and using SEAD efficiently.

13. Carl von Clausewitz, *On War,* ed. and trans. Michael Howard and Peter Paret (Princeton, N.J.: Princeton University Press, 1984), 357–67, 523–26, 528–32, 611–13. Clausewitz's views on offensive and defensive war-fighting concepts are spread throughout his book. While there is evidence Clausewitz was influenced by Hegelian views of history, he did not at that time view defensive and offensive aspects of war as aspects of a continuum or a thesis-antithesis-synthesis process.

Chapter 3

Service Approaches to SEAD

Service culture and doctrine play a significant role in the formation of single-service SEAD strategy and tactics. By understanding how the interplay of service culture and doctrine has influenced each service's evolution of SEAD, we can gain a clearer appreciation of SEAD's growing importance in the prosecution of modern war strategy and tactics. This understanding will also provide insight into opportunities for and problems with the growth of future JSEAD programs.

An analysis of each US military service's specific SEAD strategies, tactics, and procedures will help determine which SEAD elements of each service are most relevant to conducting wars in the future. How each service characterizes the threat through the lens of its doctrine and service culture is key to understanding how each employs SEAD.

As retired Air Force general Michael Dugan pointed out, the air/sea/land (and now space) roles define the basic operational turf of the services, and from these roles have developed the basic fabric of the Air Force, Navy, Marines, and Army.[1] With the creation of the institutions came the creation of the service cultures that supported them and the paradigms that charged their individual service view of the strategic and tactical environment. The parochialisms that often characterize individual service advocacy stem from this growth of service culture as well, and individual service applications of SEAD are both an example of this phenomenon and a bridge to future opportunities for true joint focuses on objectives common to all.

It is a curious fact that while each service seems obliged to advocate the primacy of its role over others, the services have each acquired elements of the others in their evolutionary processes. The Navy acquired its own "army" in the Marines and has its own "air force." The Army has its own "air force" (helicopters and special purpose aircraft) and a limited number of amphibious vehicles. Even the Air Force has made inroads into using the naval medium with its Cobra Dane-shipborne missile tracking system and has its own "army" via specially trained security police units and special forces. The key element here is not that these particular "services-in-miniature" exist but that they exist to enable their parent services to perform the tasks most directly related to the use of their medium.

SEAD is a special case in this respect because each of the services has evolved an elaborate use of the air medium in the prosecution of its war-fighting goals. Each service has a real, internal stake in employing SEAD beyond the achievement of overarching joint objectives.

Navy SEAD

We will begin with the pertinent characteristics of Navy strategy. The following is a compilation of characteristics gleaned from NWP-1, *Strategic Concepts of the US Navy* (the naval warfare publication that is their closest equivalent to written doctrine), observations made by scholars, and Navy tactics publications.[2]

Characteristics of the Navy

Missions:

- sea control
- power projection
- strategic sea lift

Command and Control:

- decentralized
- independent

War-fighting Mentality:

- decentralized execution
- wars fought as series of battles
- primarily, a raid mentality
- secondarily, a limited campaign mentality
- long-term, continuously waged war unsuited for battles of short duration due to limited storage capacity for materiel and arms and vulnerability
- "power-projection" function of specific, limited objectives

Doctrine/Tactics:

- informal, very general doctrine, links between tactics and doctrine tenuous
- tactics formulated on case-by-case basis
- lack of formal doctrine enables flexible tactics
- lack of formal doctrine frustrates strategists
- tactics vary widely; no apparent framing principles other than accomplishing specifics of mission

Threats:

- sea threat:
  - viewed primarily as threat to the fleet
  - defensive strategy accentuated because of nature of medium (i.e., destruction of ship means survival in hostile medium of ocean)
- power-projection threat:
  - target and raid specific
  - concerned with specific area of attack
  - overall land IADS not primary focus
  - use of "rollback"[3] (sequential destruction of SEAD targets) concept, but only within area of specific interest

The Navy perceives a dilemma in the function of SEAD as manifested in the disposition/employment of its air power in combat. The dilemma is a result of having to defend the fleet first before aerial power projection can be accomplished. Defense absorbs a considerable amount of force structure prior to allotting forces for power projection and makes the ratio of support/defensive aircraft to strike aircraft much higher than an "equivalent" Air Force unit.[4]

When one considers the vulnerability of naval craft without dedicated protection, it is easily understandable why the Navy places such emphasis on fleet protection. The destruction of a carrier has more impact on naval aircraft than the destruction of a particular airfield has on land-based aircraft.

A second feature of this dilemma is manifested in the way naval SEAD equipment is designed. Naval SEAD aircraft and weapons are multipurpose, whereas certain aspects of SEAD require very specific treatment. A direct comparison of the EF-111, F-4G, and EC-130H aircraft to their "equivalents" in the Navy is unfair largely because of this facet. To compensate for lack of specificity, the Navy merges a variety of tactics and equipment.

Navy EA-6s, for instance, can perform early warning/ acquisition radar and communication jamming, and target-tracking radar jamming and can fire the high-speed antiradiation missile (the AGM-88 HARM)—a task which normally requires three separate kinds of Air Force aircraft. Navy fighters and attack aircraft (with the exception of the F-14) can fire the HARM as well as deliver other more conventional standoff ordnance. Naval ships can fire the TLAM-C (Tomahawk land attack missile-conventional) at known SEAD targets with great accuracy and devastating effect. The difficulty is that, prior to the strike, friendly forces may not know the key locations of enemy emitters and fire-control nodes.

Navy aircraft that employ the HARM, however, lack the specialized traits of the F-4G, which can detect, identify, and locate very precisely its target, whether its location is known (or suspected) ahead of time or not. The Navy compensates for this by salvo-firing HARMs "in the blind," using enough missiles to "blanket" the prospective threat area. At the same time, Navy attack aircraft fly through the SAM belt to the target. This tactic/equipment meld, which stresses versatility, presumably allows the Navy to achieve maximum use of aircraft while minimizing exposure to enemy fire.

SEAD tactics for the Navy are not focused on specific radar targets; they are focused on "corridor clearing" to open a path for attacking aircraft or specific target-area suppression for a very limited geographic area. The destruction of the radar site is not the object of Navy SEAD tactics; the object, rather, is usually to suppress the radar for the period of time it takes the attacking aircraft to fly through the lethal envelope of the radar-directed weapon system en route to their targets. Because it must use nonspecific tactics and weapons (due to the multiple-duty nature of its aircraft), the Navy is more closely aligned with the piecemeal portion of the "piecemeal/integrated" continuum.

These piecemeal tactics serve the needs of Navy SEAD so long as the objectives of military power are specific, of short duration, and do not require resources needed for a long-term effort. Piecemeal tactics are suited for raids and battles, not campaigns and wars of lengthy duration that require integration with a larger combined force.

The Navy raid on Benghazi, Libya, during Operation El Dorado Canyon and some of the attacks into southern Iraq during Operation Desert Storm portray both the strengths and weaknesses of the piecemeal approach. The raids on Benghazi were extremely effective largely as a result of Navy SEAD tactics. The attacks were brief and specific, and the use of "blanket" tactics in a small geographic area overwhelmed the very surprised Libyan defenses. The criteria of specific objectives, short duration, and limited force were met. During Desert Storm, however, piecemeal tactics only worked for areas that were isolated from the main area of battle. Prosecution of single-minded piecemeal tactics in the overall campaign, while effective for the short term, was not an efficient use of SEAD resources. An integrated approach to shutting down the entire system may have required fewer sorties to accomplish the objective and, consequently, required much less exposure to hostile fire.

Using the defensive/offensive matrix as an analysis tool to ascertain Navy SEAD effectiveness yields mixed results. The prevailing philosophy of the Navy is to protect the fleet first, then project power. SEAD, as part of power projection, is used as a defensive measure to protect attacking aircraft (even though the attacking aircraft are performing an offensive role). The SEAD role, therefore, could be categorized as "defensive offense." This would limit the Navy's capability in that its SEAD forces would have to be split between protecting the fleet and protecting the attackers in the power-projection role.

As the credible threat to the US naval carrier groups diminishes, JSEAD planners are likely to see a major shift in emphasis into the power-projection category with concomitant shifts to offensive-offensive tactics. We may be seeing signs of this already. Rear Adm Riley D. Mixson, who commanded Carrier Group Two during Desert Storm, in comments written for a magazine article, detailed five areas where he felt the Navy could do better. Three of the five were directly concerned with power-projection capabilities: (1) the acquisition of longer-range attack aircraft (or tankers), (2) the need to "revise our thinking on ordnance requirements" with a focus on acquisition of more powerful, hardened-target weapons such as the I-2000 laser-guided bomb, and (3) the admonishment that the "Navy should be more aggressive in attending to the J-factor—as in *joint* warfare."[5] If the Navy is to focus more squarely on air power as a key feature of its force employment in power projection, then the next logical step is to adopt an offensive-offensive posture in achieving SEAD objectives.

It is equally likely that, as a function of allocating a greater portion of Navy resources to power projection, both naval equipment and strategies will grow closer to the Air Force's "global reach, global power" concepts and the Army's AirLand Battle-future (ABLF) concept of "the extended battlefield."[6] Both these concepts stress rapid response and flexibility and acknowledge air power as a key ingredient in shaping the three-dimensional, nonlinear battlefield of the future. The Navy's transition, in this sense, is a logical response to the evolving world environment and the military's role within it. The implications for integrated use of single-service SEAD are clear: Navy SEAD will change to accommodate *joint* SEAD.

Air Force SEAD

Over the last three years, the Air Force's analysis of SEAD has grown from one of examining single-system, enemy capabilities to assessing the damage a total enemy IADS can bring to bear on an attacking friendly force. The transition from single-system to holistic analysis of the SEAD threat was easier for the Air Force than for its sister services because the Air Force has always dealt with the concepts of three-dimensional space and time as corollaries to its strategy. The Air Force, being more technology-driven than its sister services, has tended to be more responsive to technological change. The Air Force was the first to adapt digitally reprogrammable electronic combat (EC) equipment and to recognize that integrating electronic combat into tactical thought does not occur without instituting practice in exercises like Green Flag and Cope Thunder, which stress integrating EC and SEAD into day-to-day tactical training. However, the proclivity of the Air Force to change rapidly— maybe too rapidly—and adapt to its perceived new environment has some negative aspects where SEAD effectiveness is concerned.

The primary negative effect of the extremely rapid technological changes brought about by SEAD across an institution as large as the Air Force has been asymmetric evolution: technically based agencies grew so fast that linkages between them and other less technically oriented structures were weakened, perhaps even shattered. The overall tactical structure has had difficulty adapting to the changes created by Air Force SEAD; tactical equilibrium was disrupted. This created a paradoxical situation: in order to adapt, Air Force SEAD institutionalized change itself as the primary control mechanism for achieving equilibrium. As soon as a change in the enemy IADS order of battle was detected, Air Force decision makers mandated the creation of newer pieces of hardware to "meet the threat." The mandate to build technology to counter the threat became an unchallenged axiom that created disconnects among many Air Force tactics communities that resulted from "level-of-technology" imbalances.

The spectrum of evolutionary development caused by rapid technological change in SEAD was uneven; the criteria for measuring effectiveness in one area of growth may not have applied to another. For instance, measuring bombing effectiveness was relatively straightforward no matter how sophisticated the bomb: it hit the target or it didn't. Such a criterion may have applied to SEAD during the use of relatively unsophisticated SEAD tactics in Vietnam, but the subsequent evolution of the IADS threat and the Air Force's response to it made such a simplistic criterion irrelevant. The criteria to evaluate SEAD's effectiveness—even in Desert Storm—are called into question because they were created for a technological and political world at a different stage of evolution.

The imbalance that technology has on the evolutionary development of Air Force SEAD forced a nonlinear pattern of growth in other areas as well. The balance between doctrine and strategy was destabilized. Doctrine and strategy development tended to be ahead of or behind technological reality, but rarely in consonance. In the absence of a bonding doctrine that tied SEAD strategy to SEAD tactics, tactical thought and application tended to become specialized, fragmented, and parochial. This caused significant schisms between users and builders at a very fundamental level. Entire engineering initiatives and programs were often implemented without reference to actual tactical or strategic needs. For example, one Air Force engineer, when asked why there were no representatives from the tactical flying community (those with engineering backgrounds and training) in his major weapons system program, responded, "I don't want to corrupt the pure engineering environment with extraneous influences."[7]

The institutionalization of change in Air Force SEAD has the ironic effect of estranging the "thinkers" and the "doers." Those charged with creating SEAD strategy and tactics, even though they sense the changing world around them, are seemingly doomed to lag behind technological reality in creating forward-looking concepts for the future. Those writing doctrine are forced into ever more visionary statements in an attempt to bond SEAD's technological reality to a view of the future far enough ahead to provide a framework that accommodates the changes taking place in the technical world. Unfortunately, in order to provide the required "leap of vision" into the future, writers of Air Force doctrine must use a paradigm of the world that may not apply when the future arrives.

Air Force-applied SEAD—as the leading edge of technical and tactical thought—is both the beneficiary and a victim of this situation. The Air Force's specialized aircraft, sophisticated tactics and methods, and community of expertise surrounding Air Force SEAD has resulted in a capability of staggering *tactical* effectiveness. Air Force SEAD tactics and methods change with each new technological capability; *doctrine* and *strategy* formulation have difficulty making the appropriate adjustments. Hence, doctrine and strategy are in danger of being driven by SEAD technology rather than *defining* the turf where technology is harnessed to achieve overall objectives.

The most recent pronouncement of written Air Force doctrine, the March 1992 Air Force Manual 1-1, *Basic Aerospace Doctrine of the United States Air Force,* postulates a three-dimensional application of power using speed (as a function of time), flexibility, range, accuracy (a combination of some specific "principles of war," namely, objective, mass, and economy of force), and firepower as primary elements. The manual deals with the uncertain world environment and with rapidly changing technologies by crafting a vision of air power's future and linking it to past doctrinal thought. While such visionary doctrine is a usable construct for developing strategy at the macro level, it is likely to draw a cynical response from the SEAD tactician at the "nuts-and-bolts" tactical level. The link between strategy and tactics, implied though it is, is tenuous because of the allowances that must be made for technological change.[8]

To frame the positive and negative aspects of Air Force-applied SEAD, it is necessary to detail the relevant characteristics of the Air Force in much the same way as was done with the Navy.

Characteristics of the Air Force

Missions:

- aerospace control
  - – power projection
  - – airborne defense against air attack
- aerial support of ground forces

Command and Control:

- centralized
- strong emphasis on preserving command and control
- (C^2) structures indicates relative dependency

War-fighting Mentality:

- decentralized execution geographically; centralized execution by function
- wars fought as function of objective; duration function of logistics, sustainability, and attrition
- prefer use of overwhelming force, decisive force, and sufficient force respectively[9]
- offensive: seize the initiative; exploit speed, range, flexibility of air power
- bold in planning, cautious in execution
- global power projection supporting full range of objectives from very general to very specific

Doctrine/Tactics:

- formal, general doctrine, visionary and complex; not well understood by rank and file
- very specific tactics that tend toward dogma
- formal doctrine often ignored at tactical level
  - – doctrine not part of tactical education
  - – perception exists that doctrine doesn't apply
- written doctrine often lags tactical development because of technical change; attempts to "universalize" unchanging principles; removes doctrine from tactical use
  - – dynamic relationship between doctrine and tactics idealized
- tactics support both specific target and theater objectives. Pragmatic link between doctrine and tactics implicit, not explicit strategy holistic; sees war objectives "all at once"

Threats:

- direct threats are to airborne assets; viewed as function of power projection (tactical offensive)
- protection of airfields important, but since primary war-fighting impetus on power projection, protection of airborne assets accentuated
- indirect threats to airborne assets (early warning systems, C^3, importance of nonlethal elements of IADS to enemy scheme of battle) considered as a function of campaign/war objectives

The dilemmas which plague the Air Force are more complicated than those that face its sister services both because the Air Force views SEAD differently and because of the impacts of changing technology. The Air Force defines SEAD as a part of electronic combat—unlike the Army and Navy, which generally subscribe to the joint definition and categorize SEAD as separate from command, control, and communications countermeasures (C^3CM) and electronic warfare (EW). To complicate matters further, current joint definitions do not recognize the existence of the Air Force's "umbrella" concept of electronic *combat* under which SEAD is defined. In fact, joint definitions of electronic *warfare* (which is also part of the Air Force lexicon but is defined differently) consider EW to be a subset of C^3CM.[10] These differences are presented in figure 14.

The second dilemma concerns the Air Force's perception that it must change quickly to adapt to newer technical environments.

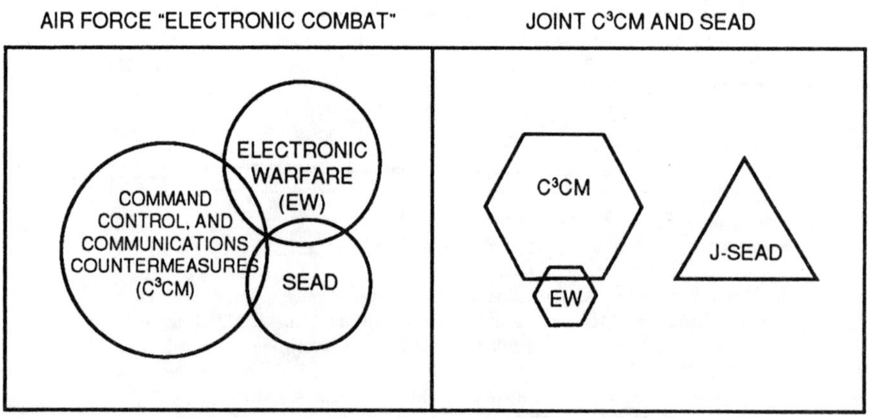

Figure 14. Structural differences.

Depicting this phenomenon is difficult; an approximation appears at figure 15.

Figure 15 shows the growth of SEAD and C^3CM, and the gradual submersion of EW. This is a reflection of the evolving nature of the threat: the distinction between the enemy's C^3 process and the weapons the C^3 apparatus directs has essentially disappeared. Therefore, SEAD and C^3CM—as far as suppression is concerned—tend to become the same thing because the target sets are increasingly the same. Only the approaches are different. C^3CM practitioners approach the target set from the perspective of degrading the enemy's ability to direct his forces; SEAD practitioners approach the target set with the view of neutralizing the enemy's ability to bring weapons to bear.

To wage a separate C^3CM campaign using the same resources that are used against SEAD targets when they are, in fact, the same targets is to waste assets. The campaign is the same, the targets are largely the same, and the weapons employed against them are the same. The Air Force structural definition seems the more appropriate choice because it (1) makes the best use of available assets for the given target sets, and (2) is more adaptable to war-fighting constructs. While there does appear to be some agreement among the services to adopt the Air Force's definition, there has been no corresponding structural change.[11]

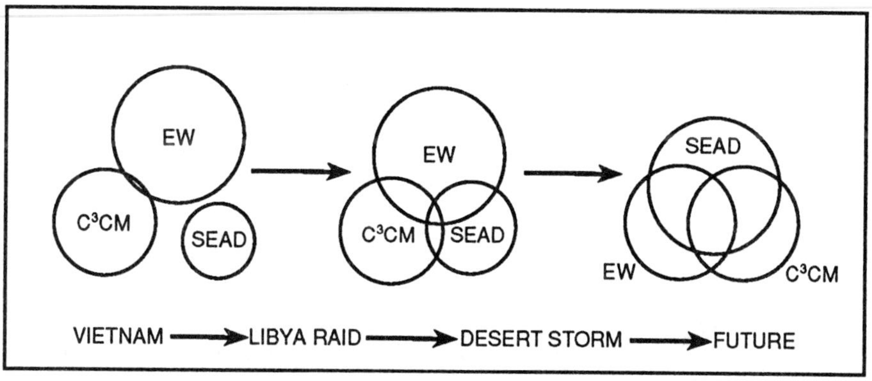

Figure 15. Effects of rapid change on SEAD.

The integrated electronic combat approach has "led the way" in the development of Air Force operational theater art. As early as 1985, with the publication of *The Tactical Air Force Guide for Integrated Electronic Combat,* the Air Force determined that as a functional part of electronic combat, SEAD can be an integrating *concept* as well as a meld of weapons systems.[12] The concept applies not only to Air Force assets but also to the SEAD assets of all services. The Air Force's "EC triad" (the F-4G, EF-111, and EC-130H) is the core of SEAD power projection. This triad represents two-thirds of the electronic combat picture and, until recently (with the retirement of more than one-half of the F-4Gs), almost all Air Force SEAD. The Air Force's EC-130H Compass Call aircraft performs the major share of C3CM, and it contributes heavily to the SEAD mission as well when it jams specific links to deny the enemy the ability to bring weapons to bear on friendly aircraft. The EF-111A Raven performs active jamming of enemy early-warning and acquisition radars to deny detection of friendly aircraft and pointing information of fire-control units. The F-4G Wild Weasel, with its ability to identify and locate specific emitters, provides lethal suppression of active radar emitters (mostly, but not limited to, fire-control radars) in either offensive or defensive roles.

The EC triad has developed into a constellation and has grown to include the airborne warning and control system (AWACS) E-3, the EC-130E airborne command and control center (ABCCC), and the RC/EC-135 (strategic electronic surveillance aircraft that can identify and locate specific emitters), as well as several other specialized aircraft.13 The Air Force-directed JSEAD constellation can also include the Navy's E-2 Hawkeye surveillance aircraft and the EP/P-3 Orion electronic surveillance and communications aircraft, as well as the EA-6B jammer and other Navy/Marine aircraft that have the HARM-shooting capability and Army assets that perform battlefield SEAD. These include the Quick Fix and Guard Rail airborne systems, which serve to detect and locate—as well as jam—selected enemy radars from airborne platforms,[14] the MLRS, and the new, extremely precise Army tactical missile system (ATACMS), along with other long-range artillery. Even the SAM-D Patriot system has been included in the extended net (because of its extremely long range) in the anti-interceptor/air defense role. The key feature in the growth of the SEAD concept, as envisioned by Air Force SEAD strategists and tacticians, is that it provides an encompassing framework in which all air power SEAD assets can be combined to achieve a theater commander's objectives.

The growth of this intricate net would not be possible using a piecemeal approach. Yet as successful as it is in executing broad objectives over the long term, an extended net may not be suited for raids and campaigns of short duration and limited objectives because of its complexity and required support infrastructure. It is neither a concept nor a system that can be set up overnight.

The EC cell that worked during Desert Storm used the Air Force concepts of integration employed over the last 10 years in various exercises. US Navy, Marine, and Army specialists augmented the Air Force Desert Storm EC cell team with their expertise, but the structure through which the team operated was generated by Air Force concepts of EC and SEAD integration. The approach seems to have been responsible for nascent developments currently taking shape for the creation of true JSEAD strategies and tactics.

The integrated SEAD approach the Air Force employed is largely a result of the Air Force view of land-based war. The Navy's war-fighting mentality is geared primarily to the raid or the short campaign as projected from the sea. The Air Force's war-fighting mentality, while desirous of a quick conclusion to warfare, prepares for wars of longer duration—especially where power projection is concerned. In this long-term calculus, the Air Force has determined that overall SEAD objectives, achieved through the integrated use of its air power resources, is the most effective avenue. Although the integration of numerous aircraft of widely differing SEAD capabilities is difficult, Air Force experience indicates it is manageable and effective for the long term.

A criterion continuum currently discussed among strategists at the macro level—as well as tacticians at the micro level—is the "threat-/capability-based" force structure spectrum. Previously, during the cold war, force planners knew who the enemy was and what capabilities the enemy possessed.

Planners developed force structures, organizations, and weapons to counter a clearly defined threat. Since SEAD planners no longer have such a clear criterion against which to measure, they have no clear way of determining what is needed. A physical threat to US aircraft, however, becomes immediately tangible when US forces are called upon to fight, regardless of whether the adversary is known. The change in planning philosophy is that strategists and tacticians no longer know how the threat will manifest itself or exactly where it will be. A capability—no matter how diffuse—must exist to counter it.

The Air Force, being firmly entrenched in "threat-based" tactics and equipment, is at a disadvantage. The nature of the future threat precludes the precise knowledge of enemy systems to which Air Force SEAD planners are accustomed. Nonetheless, even a totally capability-based force structure must be geared to a threat of some known dimension. The Air Force is moving reluctantly toward the capability-based side of the continuum but still maintains that more effort must be expended to determine the nature and content of the future SEAD threat. The Air Force also maintains that, because the future threat is tending towards highly complex, integrated weaponry netted together with increasingly sophisticated technology, emphasis on research and development for new weapons must continue to enhance the viability of our shrinking inventory. The difference between the Air Force and other services on this account is one of emphasis. The Air Force is willing to commit considerably more resources than the other services to achieve this goal.

Additionally, the Air Force is moving from a purely threat-based criteria in both strategy and tactics to one which contains an overall balanced approach. Because the Air Force's tactical heritage is largely technological, it has focused on countering the purely technical aspects of the enemy's order of battle. The evolution of many separate systems to counter separate threats was the outgrowth of this thought process.

The Air Force's general fixation on the threat has driven an acquisition process which may have overstated its case in that, with the demise of the Soviet threat, the requirement to acquire specific systems designed to counter specific Soviet threats disappears. In an environment where there is a clearly defined threat, there is justification for a commitment of resources to meet the specific threat. In an environment where the threat is diffuse, it becomes more difficult to justify a commitment of resources to develop systems designed to counter specific threats. There is a dilemma here. On the one hand, because Air Force acquisition strategies have been tied to building systems geared to defeating specific Soviet threats, SEAD systems have been specifically tailored. They have also been relatively expensive. On the other hand, the lack of a specific threat upon which to tether future acquisitions has severely curtailed the Air Force's acquisition strategy. A SEAD threat exists, but its form and nature have yet to be defined in terms clear enough to justify expenditures on the basis of the cold war paradigm. Because of this, there is a movement away from the acquisition of specific systems to a more generalized use of existing platforms to meet the diffuse, complex SEAD threat of the future.

The case of the Wild Weasel mission is illuminating in this respect. The F-4G Wild Weasel is the only aircraft in DOD that was designed to identify, locate, and destroy specific radar emitters by mating an extremely sophisticated radar homing and warning receiver to an equally sophisticated antiradiation missile. (The Weasel community uses the phrases "smart plane, smart missile"—as opposed to "dumb plane, smart missile"—to distinguish the Weasel from aircraft that can fire the HARM but that do not possess the F-4G's formidable detection and location system.) Because the F-4G represents a one-of-a-kind system, it delivers an extremely specific, technical capability that has proved extremely valuable to SEAD operations. Nonetheless, the Weasel community's single-minded focus of performing a lethal aspect of the SEAD mission militated against acquiring a general HARM-firing capability on other Air Force aircraft—until very recently. The F-16 already has acquired this capability, and the F-15E may soon acquire it. Apparently, the "dumb plane, smart missile" combination represented by the current F-16 and its naval equivalents, the F/A-18 and EA/A-6, cannot match Weasel standards. But these "dumb aircraft" do not cost as much and can be used for many other missions besides SEAD. In a resource-constrained environment, less costly multipurpose aircraft, while not as effective as the F-4G, are more attractive to the military community. While a SEAD threat clearly exists, it cannot be defined with enough precision to justify the expense of maintaining a specific-mission aircraft.

The specificity of the F-4G mission, along with the F-4G's age and cost—like the natural selection process in evolution—may have led to "selecting" the Weasel aircraft and much of its mission out of existence as a way of accomplishing SEAD in the future. This "dinosaur syndrome" appears to be the case as the Air Force evolves from threat-based to capability-based criteria.[15]

Another important example of the Air Force's move toward capability-based SEAD was the joint forces air component commander's (JFACC) decision to use the F-117 as a SEAD asset during Desert Storm. The SEAD role for the F-117 is an unconventional one, but the F-117's unique "radar invisibility" characteristic makes it ideally suited for some types of SEAD missions. By being "invisible" to radar and infrared detection, the F-117 pilot needs only know where the target is in order to attack it.16 The F-117, however, was never designed to replace the Weasel because it has no way of detecting mobile (position unknown) emitters and is totally dependent on exterior sources of information to accomplish the SEAD role. Without extensive and expensive modifications, the only essential difference between the F-117 and other non-F-4G HARM-firing aircraft is the F-117's stealth capability and accurate bombing system. With respect to SEAD targets (active emitters), the F-117 is in precisely the same position as the F-16 and the marginally better-equipped F/A-18s and EA-6Bs, which have limited capability to detect and locate emitters.

Nevertheless, the F-117 is capable of performing some extremely innovative SEAD tactics. Its stealth feature makes it an ideal platform for destruction of known strategic IADS facilities, and it very impressively demonstrated these capabilities during Desert Storm. The F-117 can make a significant contribution to the SEAD effort, but SEAD planners should consider the capabilities of the F-117 as *a* tool, not *the* tool, for destructive SEAD.

The use of separate aircraft to perform specific functions in the SEAD environment seems to be an outgrowth of the "threat-based, threat-specific" resource use of the Air Force. By using the capability and integration approaches in this continuum, the Air Force could probably make its F-15s, F-16s, and F-111s virtual stealth aircraft with extensive EF-111/EA-6 jamming. Multiple tactical deception scenarios that used both stealth and jamming to confuse the enemy's targeting calculus would degrade the enemy's ability to conduct air defense operations.

By comparison, the Navy appears always to have followed a "capability-based" approach to tactics and weapons systems acquisition for SEAD. Air Force SEAD tacticians need to open a dialogue with the Navy on this account. Navy air assets must be capable of responding simultaneously to a wide variety of threats; the luxury does not exist—either as a matter of space or budget—to have tailored forces for specific threats.[17]

Even so, there are some aspects of SEAD that cannot be accomplished effectively by multiple-role aircraft even though they are technically "capable" of achieving the objective. "Capability-based" criteria demand an effectiveness threshold that satisfies the long-term SEAD requirements of the forces employed and that has the desired level of effect on the enemy.

The Air Force stands at the far end of the offensive side of the defensive/offensive continuum. There is, perhaps, no other concept so deeply ingrained in the Air Force aviator as the concept of the strategic offensive led by air power. It is an intuitive assumption on the part of the Air Force SEAD planner to support the "air power" effort. That "effort," though not often explicitly identified, is the use of air power as the decisive element in destroying the enemy's capability and will to wage war. Air Force doctrine has never changed on this matter and never will.

The evolution of SEAD within the offense-based strategy and tactics of the Air Force, however, weaves a curious web. The initial use of SEAD was defensive. As the threat to air superiority moved from the threat posed by other aircraft to the threat posed by the integrated use of ground and air assets (air deniability), SEAD took on greater meaning for the prosecution of the superiority mission. Initially, SEAD assets proliferated piecemeal to protect specific missions against specific threats. Air Force SEAD evolved into a lone concept whereby the SEAD mission supported air superiority. Finally, Air Force SEAD seems to have evolved to the point where it is an offensive element of strategy which recognizes that a "SEAD campaign" is essential to achieving air superiority when prosecuting a war against an enemy possessing a sophisticated IADS. As this trend toward sophisticated IADS appears to increase, the offensive application of SEAD also appears to be increasing.

The Air Force, because of the encompassing manner with which it views the battlefield, tends to view SEAD as a multiple-level application of assets and tactics. The current lexicon refers to this as an "attack in depth" similar to the Navy's EC rollback concept, except that it extends to the entire IADS and mobile air defense array. The multiple-level application of SEAD manifests itself most clearly in the SEAD campaign. It is both iterative and simultaneous. This means that, while the campaign is built in "phases," these phases can occur simultaneously. The phasing occurs as a matter of emphasis. The heavier emphasis flows from one phase to another as the situation dictates. The phases of the campaign are usually set out sequentially to destroy/disrupt enemy

1. command, control, and communications,
2. early warning systems (including associated enemy air interceptors),
3. acquisition systems dedicated to fire control, and
4. fire control and air defense systems.[18]

At each point, the intent and direction of the strategy are clearly offensive in nature. The "offensive-offensive" development of Air Force SEAD is a product of the Air Force's doctrine, its emphasis on technology, its intuitive grasp of three-dimensional warfare, and its sensitivity to changing realities. On each of these accounts, Air Force SEAD differs from other services' SEAD.

The Air Force, however different it may have become in its SEAD outlook when compared to other services, is the offspring of the Army Air Force (AAF). Historically and recently, the most common bonding of services with respect to the development of SEAD has been between the Air Force and the Army.

Army SEAD

In his 1978 master's thesis for the US Army's Command and General Staff College, Air Force major James L. Hendrickson made these observations:

> The difference between U.S. Army and U.S. Air [Force] employment of electronic countermeasures (ECM) affects the doctrine, internal procedures, and organizational structures of each component service. The Army employs ECM as an offensive combat support measure much like artillery. The Air Force employs ECM primarily as an aid in penetrating enemy air defenses and protecting aircraft for maximum weapons delivery effectiveness.
>
> Therefore, Army operational units are less dependent on the deployment of ECM to operate their organic weapons systems then [sic] are Air Force airborne weapons systems. . . .[19]

While these comments apply to electronic countermeasures, they hold true for SEAD as well. Until very recently (i.e., with the acquisition of force-level numbers of Blackhawk deep-strike troop transport and the deadly firepower of the day-night-capable Apache helicopters), Army SEAD was a purely "support air power" concept viewed by many Army operatives as a drain on their combat power. A historical feature of Army SEAD was to retain control of assets used to support the JSEAD campaign and release them to the theater commander *only after* Army requirements had been met. During Desert Storm, however, the theater commander had virtual control over all assets, and—as often as not—directed Army support for Air Force SEAD operations, even at the expense of withdrawing support from some frontline US Army units to bolster the JSEAD effort.[20]

The Army's use of SEAD since the mid-1980s has evolved rapidly both as a function of its own modernization and of pressures to pursue jointness. The Goldwater-Nichols Initiative of 1986 probably had some influence on this as well. The history of Air Force/Army cooperation in SEAD, however, predates Goldwater-Nichols by at least seven years with the establishment of institutionalized links between the Army's Training and Doctrine Command (TRADOC) and the Air Force's Tactical Air Command (TAC). Two organizations, the Airland Forces Agency (ALFA) and the Center for Low Intensity

Conflict (CLIC), were the result of TAC commander Gen Wilbur Creech's efforts to form a bond between "natural allies" in the prosecution of the land war. The AirLand Battle concept so eloquently espoused in the Army's basic manual, Field Manual (FM) 100-5, *Operations,* reflects the results of General Creech's efforts.

The arrangements the Army and Air Force made are reflected in their participation in each other's larger exercises, such as the Army's Cobra Gold in Pacific Air Command, Twelfth Air Force's Sagebrush series of exercises, and TAC's Green Flag/Blue Flag exercises. ALFA produces numerous documents on Air Force/Army joint tactics in ground maneuver and combined arms with a focus on the close air support rendered directly to ground troops in contact with the enemy. With the advent of Army systems that can reach farther with great speed and accuracy, the distinction between close air support and interdiction (formerly defined by the fire support coordination line, or FSCL)[21] is no longer limited by the range of fire of the ground forces' systems. With the acquisition of extremely long-range artillery weapons—the MLRS, advanced 155-mm weapons, and the extremely accurate advanced tactical missile system, along with the mobile extension afforded by Blackhawk and Apache attack helicopters—the area in which Army CAS and interdiction aircraft operate has expanded. Additionally, because of the expanding battlefield, exposure of friendly aircraft to enemy organic air defense systems has increased. With more Army assets now exposed to the same general threat as their Air Force counterparts, the impetus to develop truly joint SEAD procedures has gained full force.

This evolution of Army battlefield strategy and tactics has caused difficulties in defining time-honored concepts such as the fire support coordination line and the "deep area" of the battlefield. With the acquisition of artillery of much longer range, warheads of greater precision, and other weapons that can operate with virtual impunity anywhere on the battlefield, it is uncertain whether any of these former terms can be used to differentiate division of labor in the AirLand Battle for combat air support and interdiction, let alone SEAD.

The Army is wrestling with the issue of the expanding battlefield and the impact of air power. The resulting Army SEAD concept is its new AirLand Battle-future concept. It extends the nonlinear battlefield where attack can be accomplished nonsequentially based on time-sensitive objectives. Previous battlefield concepts of attack were based primarily on geographic objectives. Oddly, the newer AirLand Battle-future concept (vice AirLand Battle) appears to be an adaptation of the Air Force's three-dimensional concept of the battlefield expressed in the Army's terms.

During Desert Storm, the standard Air Force air tasking order (ATO) identifying targets beyond the FSCL were often considered interdiction sorties when, in fact, they may have been in support of strategic targets, SEAD, or other related functions. Both the Air Force and Army suffered the same difficulty identifying where interdiction stopped and battlefield SEAD began. Army SEAD is put in perspective when we detail the Army's general characteristics.

Characteristics of the Army

Missions:

–"deterrence through readiness"[22]
– –prepared to fight high-, medium-, and low-intensity conflicts
–land control
– –defend the homeland
– –defeat the enemy in battle

Command and Control:

–centralized, but fully capable of independent action
–adherence to plan critical

War-fighting Mentality:

–strong preference for the offensive, but heavy emphasis placed on defensive aspects of war
– –primary tenets: initiative, agility, depth, synchronization
– –separates battlefield into close, deep, and rear. AirLand Battle-future changes this to a more fluid concept of "the extended battlefield" in which this area is defined more as a function of time and purpose than location[23]
–conditioned to fight wars "over the long haul," but constituted to fight rapid, combined arms, maneuver warfare to achieve rapid victory
–dependent on huge logistics infrastructure requiring emphasis on defensive strategies and tactics

Doctrine/Tactics:

–formal doctrine, very specific, and well understood throughout force; pragmatic links to tactics both implicit and explicit
–tactics and strategies clearly driven by doctrine; extremely strong relationship among doctrine, command, execution of plan, and tactics
–heavy dependence on doctrine as reference for tactical operations—even in novel situations
–tactical flexibility restricted by adherence to formal doctrine
–strategic planning enhanced by doctrinal clarity and clear relationships between plan and ability to execute

Threats:

–complex; vulnerable to air attack and heavy artillery; tactical defeat through being outmaneuvered (positional advantage)
– –offensive: destroy the enemy's ability to inflict damage
– –defensive: retain force enough to slow enemy advance by gradual attrition of enemy forces, making the price too dear to pay; train forces to absorb offensive blows by practicing "ability to operate in survival situations"
–threat viewed both strategically and tactically as function of "high-to-low" level of conflict
–destruction of IADS per se not major focus; concerned with SEAD as function of immediate battle objectives

Modern Army SEAD is more a tactical concept than a set of specific weapons and procedures. Artillery, special forces, heliborne assault, or electronic jamming are used to perform the suppression function. The distinction between JSEAD and SEAD, for the Army, is largely a function of who does what for whom rather than what is done. If the Army performs the function for itself, it is SEAD; if it performs the function in support of another service or at the direction of a joint commander, it is JSEAD. The same holds true if the function is performed by another service for the Army.

A comparison of Army, Navy, Air Force, and Marine SEAD tactics reveals that the Army is closer to the Navy in its piecemeal approach to the use of SEAD—not because of a raid mentality, but because the Army is not tactically constituted to see the total war area "all at once" in the same way air power users might see it. The Army, being restricted primarily to the ground, is constrained to think of territory and miles gained per day. The functional frame of reference still appears to be denominated in "chunks of land" rather than the enemy's ability to fight over a period of time. Therefore, meeting SEAD objectives tends to be seen as a function of acquiring specific territory.

The piecemeal approach suits the Army's purpose as well as the Air Force's in areas where the SEAD target is of critical interest to both. It is only in situations where other service assets are not available to perform SEAD against a target and SEAD does not suit the immediate objectives of the Army's plan that difficulties along the piecemeal/integrated continuum occur.

Army SEAD on the defensive/offensive continuum shows a balance in spite of the strong offensive overtones of FM 100-5. The increasingly complex threats the Army faces have sensitized the Army's view of SEAD. The threat the Army faces may consist of both enemy field-army air defenses and elements of the enemy's strategic IADS. To this threat may be added enemy battlefield air interdiction aircraft as well as the entire array of enemy artillery and ground troops. When SEAD gives the Army relatively uninterrupted use of the ground because of localized air superiority brought about by SEAD actions, it is a positive element in Army offensive strategy. When SEAD defends forces so that Army elements may "fight another day" (whether or not they are engaged in offensive actions), it is a positive element in the Army's defensive strategy. Again, the balance is dictated by the scope of the battlefield rather than the overall objectives of the battle itself.

Using the resource continuum shows that the Army, being very deliberate in its processes as a function of doctrinal adherence, ensures that it has what it needs before entering the battle. This is also a function of the "initiative, agility, depth, and synchronization" tenet scheme that characterizes AirLand Battle doctrine. One does not seize the initiative without the required tools or a well-thought-out plan of execution.

> A detailed, comprehensive JSEAD plan is of little value if it is poorly executed. The planning step is but half the requirement. On tomorrow's battlefield, combat plans will require aggressive, synchronized execution if we are to accomplish the mission and inflict our will upon the enemy. An operation's success must be "rapid, unpredictable, violent and disorienting. . . ." More than any other factor, synchronization of forces is an operational necessity in the execution of JSEAD.[24]

The Army can bring to bear significant SEAD resources, though they are composed differently than those of the Air Force and Navy. The Army possesses no antiradiation missiles, though it does possess a detection and location system that can provide the general locale of a threat emitter. The Army also possesses a formidable capability to insert special forces to accomplish numerous tasks—support for the SEAD effort being one of them. The first SEAD action of Desert Shield/Desert Storm was performed with the aid of Army helicopters. The Army has artillery that is extremely accurate to ranges exceeding 30 kilometers (sufficient to destroy enemy air defense systems that are threats to friendly CAS/battlefield air interdiction sorties, providing the enemy system's exact location is known).

The Army's capabilities supporting the AirLand Battle future are largely in place or have been funded. These forces promise to provide a much wider range of tactical options and opportunities for joint tactics development in JSEAD. Table 8 shows some of these weapons.[25]

Additionally, there is strong congressional pressure on the Army to acquire a ground-launched version of the Tacit Rainbow radar harassment weapon. This system is designed to loiter over the radar elements of an enemy's IADS (battlefield or strategic), wait for a radar to emit, detect it, locate it, and attack it. The air-launched version of this system was cancelled by the Air Force in 1991 because of technical and contractual difficulties, but the requirement for a system that performs the Tacit Rainbow's functions still exists.

The addition of a radar harassment system such as the Tacit Rainbow to any of the service's SEAD inventories—especially the Army's and Air Force's—complicates the enemy's defense calculus enormously, both in real terms and psychologically. The Tacit Rainbow drone was designed to be launched in large numbers to saturate specific areas of the enemy's defensive array. Since it loiters for long periods over its intended victims—waiting for them to radiate—it obligates the enemy site operator to decide whether to use radar. An enemy site operator who elects to radiate becomes a likely candidate for targeting by the drone. An enemy site operator

Table 8

AirLand Battle-Future Key Systems

FIELD ARTILLERY	OTHERS
ADVANCED FIELD ARTILLERY SYSTEM/FUTURE ARMORED RESUPPLY VEHICLE	JOINT SURVEILLANCE TARGET ATTACK RADAR SYSTEM
MULTIPLE LAUNCH ROCKET SYSTEM	UNMANNED AERIAL VEHICLE
TERMINALLY GUIDED PROJECTILE	COMMON GROUND STATION
TERMINALLY GUIDED WARHEAD	ENHANCED POSITION LOCATION REPORTING SYSTEM
SEARCH & DESTROY ARMOR	MOBILE SUBSCRIBER EQUIPMENT
ARMY TACTICAL MISSILE SYSTEM	IMPROVED HIGH FREQUENCY RADIO
GROUND-LAUNCHED TACIT RAINBOW	ATTACK HELICOPTER
FIREFINDER	

who does not radiate is effectively neutralized for the period the radar remains off.

The Air Force also has the opportunity to integrate Army concepts into its SEAD strategy. There is congressional pressure on the Air Force to field a ground-launched version of the Tomahawk (conventional warhead ground-launched cruise missile), and this offers considerable opportunity to integrate manpower and concepts between the Army and Air Force in bilateral SEAD operations.[26]

Army SEAD seems to fall in the middle of the threat-/capability-based continuum. The application of artillery is generic, and, so long as the enemy site can be identified and located, the use of weaponry is not tailored to the specific threat. Communication and radar jammers can be tailored to a specific threat array, but forces and tactics are not driven by a jammer's characteristics. The Army order of battle is designed such that its tactics and weapons can be tailored to meet a specific threat while remaining generic enough to apply to most situations.

The key to understanding Army SEAD is in the perception of what is at stake. As Army assets become more and more vulnerable to elements of an enemy's IADS array, the Army becomes increasingly sensitive to protecting its investment. The primary impetus of AirLand Battle future, however, is that the extended battlefield shifts the formerly indirect consequences of SEAD action as performed by other services into the direct-consequences category because of the speed of battle and the nonlinearity of the battlefield. The "new reality" promises to give added incentive in hastening the conversion to a truly joint application of SEAD tactics and weapons. Table 9 shows the differences between the current AirLand Battle concepts and the proposed concept of AirLand Battle-future/ AirLand Battle operations contained in TRADOC Pamphlet 525-5, *AirLand Operations: A Concept for the Evolution of AirLand Battle for the Strategic Army of the 1990s and Beyond.*[27] The changes are dramatic. The concepts of flexibility and speed have transformed the Army's view of the battlefield and the role SEAD will play upon it.

The distinctions between Army uses of the electromagnetic medium and those of the Air Force are also deeply affected by the movement towards true JSEAD. Army electronic warfare had focused principally on C^3CM. Army SEAD was primarily the application of artillery and rockets to destroy identified enemy air defense positions in the forward edge of the battle area. Like the Air Force, the Army's ability to make clear distinctions between C^3CM and SEAD is diminishing. A more holistic process has evolved which views the problems posed by the enemy's air defense apparatus in the context of the expanded battlefield.

As the battlefield distinctions separating near from far and weapon from weapon controller become more diffuse, so do the definitions of C^3CM and SEAD. In this sense, US Army electronic warfare and SEAD, though officially subscribing to the separation mandated by joint definitions, are evolving to a functional use of electronic combat as currently defined by the Air Force.

Table 9

Differences between AirLand Battle and AirLand Operations

DECADE OF CHANGE - CHALLENGE FOR THE 1990'S

ALB VERSUS ALB-F/AIRLAND OPERATIONS

Key Points	AirLand Battle	AirLand Battle - Future/AirLand Operations
Orientation	Joint & Combined Warfighting	Joint, Combined and Interagency Operations across the Operational Continuum
Levels of War	Tactical/Operational Focus	Tactical Insights, Operational Focus, Operational to Strategic Link
Mission Focus	Forward Deployed/Reinforcing Mature Theater	Force Projection (from CONUS/OCONUS) - Deploy then Employ Immature Theater (possible forced entry)
Threat Array	Soviet-style Echeloned Forces Linear Superior Numbers	Region Specific Linear/Nonlinear Approximate Force Parity
Battle Focus	Defensive Close-Deep-Rear Operations	Offensive One Extended Battle
Task Organization	Standard Corps	Tailored Corps (METT - T Dependent)
CSS	Reactive Supply Point Distribution	Anticipatory Unit Distribution (with tailored, agile CSS forces)
Tenets	Initiative, Agility Depth Synchronization	Increased Premium on Initiative and Agility Increased Depth (Extended battlefield) Critical Synchronization (More a complex)

TRAINING AND DOCTRINE COMMAND

Source: TRADOC Pamphlet 525-5, "AirLand Operation: A Concept for the Evolution of AirLand Battle for the Strategic Army of the 1990s and Beyond."

Marine SEAD

Marine doctrine and tactics present an ironic set of circumstances with respect to SEAD. On the one hand, the Marines operate routinely as a combined arms force integrating their land, sea, and air assets. They adhere to clear doctrine that is comprehensible to the rank and file and use it often during exercises and in real-world situations. The doctrine is reflected at every point in their strategy and tactics; application of combined arms is consonant from overall doctrinal theory to specific weapons' application use by the smallest tactical unit. Marine doctrine and employment concepts represent a microcosm of "jointness" within the confines of their combined operations.

On the other hand, Marine Corps use of SEAD can be construed as parochial and very limited when compared to other services'.[28] The SEAD resources the Marines use are limited, and the SEAD objectives that define their tasks are designed primarily to support Marine-controlled operations. The paradox is that, even though Marine SEAD is consonant within the "mini-joint" coordination involved in its combined arms operations—it is two-dimensional. Marine SEAD strategy uses the three-dimensional capability of Marine SEAD to support two-dimensional land warfare objectives. As such, it is considered a lower level tactical support element. The Marines' primary doctrinal manuals don't even mention SEAD as a concept, and the term is not contained within the glossary of Fleet Marine Field Manual (FMFM) 1-2, *The Role of the Marine Corps in the National Defense*.[29]

Nonetheless, Marine Corps SEAD does exist—not at the doctrinal level but at the tactical level, where it serves as a tool in the achievement of combined arms operations. Marine doctrine's staunch emphasis on using combined arms maneuver warfare to accomplish land warfare objectives creates difficulty for integrating Marine SEAD with more encompassing concepts that accentuate the three-dimensional qualities of SEAD in achieving overall war objectives.

Marine military thought is tactical and is based primarily on the concept of maneuver. The Marine Corps' scheme of battle relies on the use of combined arms to "shape the battlefield," and it is through these lenses that Marine tacticians view SEAD.

> Marines are generally most familiar and comfortable with the tactical realm of war, which is concerned with defeating an enemy force at a specific time and place. The tactical level of war is the world of combat. The means of tactics are the various components of combat power at our disposal. Its ways are the concepts by which we apply that combat

> power against our adversary. These are sometimes themselves called tactics—in our case, tactics founded on maneuver. Its end is victory: defeating the enemy force opposing us. In this respect, we can view tactics as the discipline of winning battles and engagements.[30]

The US Marines provide a unique function among the services. One of the primary roles that distinguish the Marines was pointed out by a former ambassador to the Soviet Union who noted, "The Marines are a kind of military crisis action team for the Department of State."[31] Though his words were no doubt intended as a quip, the ambassador struck at the core of the difference between the Marines and all other services: the Marines are *required*—as a matter of mission—to respond quickly to any contingency with the resources they have at the time. "Ultimately, the Marine Corps' *true mission* must meet the recognized needs of the nation while ensuring flexibility to adjust for unanticipated requirements."[32] Meeting requirements has led the Marines to develop "tactics as doctrine" and to acquire forces that can combine all elements needed to project power quickly, flexibly, and decisively in the pursuit of specific goals. Therefore, it is not peculiar that such a specific application of power as SEAD would not be mentioned in doctrinal documents on warfighting, campaigning, and roles.

However, SEAD, as a tactical concept, *is* addressed in support of specific Marine objectives, but not as an objective itself. Marine EA-6B and F/A-18 aircraft employ some of the most sophisticated SEAD target area tactics in the air power community, but these tactics are developed conceptually as supporting measures for Marine operations. As there is no framing Marine doctrinal concept exterior to Marine war-fighting concepts, the use of Marine SEAD assets exterior to Marine operations has no doctrinal basis: SEAD is a direct outcome of what the air tasking order dictates (target, time-on-target, general nature of task).

Though Marine SEAD concepts derive primarily as a result of supporting Marine operations, there is considerable discussion in their professional journals about what other services are doing doctrinally and tactically. One of the most illuminating articles, in fact, on the Army's new AirLand Battle-future concept surfaced in the *Marine Corps Gazette*.[33]

Marine tactics publications on SEAD, mostly classified, display a keen awareness of what other services think and do. On the surface, the Marines would seem to be more open than other services to ideas of jointness because of the framework of their combined arms scheme of battle, but, curiously, this is not the case.

Precisely because the Marines perform their combined arms scheme of battle so well, they tend to be cautious of adopting joint procedures that are not nearly as well developed nor integrated as those already employed internally. While the Marines display a superb ability to combine the elements of their land, sea, and air assets into a truly cohesive fighting force, it is difficult to divest separate elements of their SEAD assets as an integral part of a "joint force." Desert Storm showed that it is likely to take some time for the Marines to embrace jointness as a preferred method of SEAD employment.[34]

Marine SEAD tactics present an irony. Though these tactics are not mentioned as part of an overall campaign element, when they *are* written they are so specific they tend to be classified and out of the mainstream of tactical thought. The irony is that Marine SEAD is a topic not "important" enough to be a part of published war-fighting doctrine, yet it is "too important" as a tactic to be public knowledge.

Marine SEAD tactics at the "nuts-and-bolts" level are mostly a blend of Air Force, Army, and Navy approaches to employing air power on the battlefield, with some unique Marine twists. The following synopsis of Marine service culture, doctrine, tactics, and salient viewpoints helps explain why the Marines perceive SEAD as they do.

Characteristics of Marine SEAD

Mission:

- flexible application of military power as directed and required to meet specific objectives through integrated use of land, sea, and air assets
  - in association with the fleet, meet worldwide power-projection needs in
  - peace and war readiness for expeditionary service
  - reliable performance[35]

Command and Control:

- centralized command; stresses single commander in charge of war-fighting operations
- execution can be centralized and decentralized depending on the situation

War-fighting Mentality:

- aggressive, straightforward; specific-task oriented by geography or function
- can be defensive or offensive based on direction of higher authority
- highly disciplined
- offensively constituted to achieve objective quickly and occupy for as long as required
- defensively constituted to "hold the course" for as long as required
- trained to improvise both tactically and strategically given limited resources

Doctrine/Tactics:

- formal, specific, pragmatic; required to be, and is, well understood by rank and file; tactics are direct outgrowth of doctrine; significant consonance among doctrine, strategy, and tactics
- doctrine appears to enable flexible use of tactics within confines of Marine use; ground warfare tactics heavily influenced by Army; air tactics heavily influenced by both Navy and Air Force; sole owner of amphibious tactics
- heavy allusion to historical traditions; very slow to change fundamental doctrine
- reluctant to adopt technological changes unless they fit into overall doctrinal scheme
- "Joint operations are a means to an end—not the end in itself. Joint operations in their larger application are essential when one strategic task requires the coordinated employment of continental and maritime operations."[36]
- air power is specific and integrated as a function of combined arms activity; it is subsumed under maneuver warfare

Threat:

- most complex of all services. Threat perceived to entire force as function of overall objectives.
- Forces adapt to land, sea, air threat as situation dictates
- SEAD is a function of protecting forces in area of Marine interest so they may prosecute the objective (a specific place or piece of territory at, for, or by a particular time)
- applies generic tactics and weapons to suppress the threat as doctrine demands global flexibility and versatility

Marine SEAD theory is tactical, pragmatic, and designed primarily to be used in conjunction with Marine operations. The FMFM 1-2 reference to "continental and maritime" operations shows that Marines view joint warfare primarily as a function of two-dimensional warfare. The phrase "continental and maritime," as opposed to land and sea, is an allusion to traditional constitutional language. In an attempt to bind Marine doctrine to historical tradition, Marine doctrine may have slighted the reality of the three-dimensional twentieth century. Marine SEAD, therefore, is designed to support land or sea operations from the air as an extension of artillery and troop cover on the two-dimensional battlefield. An analysis of the effectiveness of Marine SEAD tactics and strategies using the four continuums discussed in chapter 2 reveals much about general Marine applications of air power and its relation to SEAD.

The need-/resource-based continuum, both as a function of doctrine and historical record, shows that the Marines operate from the limited-resource basis. Their SEAD tactics are driven largely by the availability of resources. The emphasis is on achieving the objective with assets at hand; as a result, Marine tactics tend to be very creative given the limited resources with which they often must work.

Directly related to the need-/resource-based spectrum is the threat-/capability-based continuum. Largely because the Marines have close affiliation with the Navy and because their doctrinal tradition dictates it, Marine SEAD is heavily capability-based. They have neither the resources nor the threat-specific mentality of the Air Force, or, to a lesser degree, the Army. Therefore, to accomplish SEAD, they tend toward a meld of tactics and weapons matching the best overall capability to achieve battlefield objectives.

With respect to the defensive/offensive continuum, the Marines—like the Army—exhibit strong characteristics at both ends of the spectrum because the dynamics of battle often require both capabilities. The Marines, because they are usually the first land forces available to prosecute American foreign policy interests, are often directed to perform purely defensive roles even though their doctrine shows a strong preference for seizing the initiative. Marine SEAD's primary function is protecting aircraft, not projecting power in and of itself.

On the piecemeal/integrated tactics continuum Marine SEAD exhibits elements of both extremes. In application, Marine SEAD is target- and area-related and, thus, tends to be piecemeal in its application. This tendency is especially noticeable when Marine SEAD is employed to support other services while operating under Marine authority, but much less so when operating as an integral part of a Marine-conducted operation.

JSEAD, as both a composite of different services' doctrinal views, strategies, tactics, and equipment, and as a concept greater than the sum of its parts, is deeply influenced by the respective service cultures. Some war-fighting situations clearly call for the application of one service's war-fighting view over another's. A joint commander's JSEAD formulators on the JFACC staff (usually the EC planning cell) must be thoroughly familiar with each service's way of doing business in order to prosecute a "joint" war. As Desert Storm demonstrated, we have come a long way toward a "true" joint application of JSEAD, but we still have a long way to go.

Notes

1. Gen Michael J. Dugan, USAF, Retired, "Air Power Concentration, Responsiveness, and the Operational Art," *Military Review*, July 1989, 21.

2. As noted in chapter 1, the US Navy has no equivalent to the Air Force's AFM 1-1, the Army's FM 100-5, or the Marines FMFM 1- series of doctrine documents. The most recent iteration of their closest equivalent, NWP 1 (rev. A), *Strategic Concepts of the US Navy*, is dated May 1978 and reflects an extremely flexible, independent nature. Comments made by two Navy admirals speaking under nonattribution rules in January 1992 indicate this is likely to change soon. The new Navy doctrine will continue to be flexible, but not nearly so independent. Navy SEAD doctrine, per se, does not exist, and what can be said of it must be extrapolated from context. The interpretation thus rendered is deeply subjective.

3. Both Navy and Air Force SEAD tacticians use the term "roll back," but they mean significantly different things by it. The Navy sees "EC rollback" primarily as a defensive measure to protect its attacking aircraft. The Air Force concept of "EC rollback," contained in the *Tactical Air Force Integrated EC Guide*, reflects an offensive-oriented use of EC "layering"—the coordinated efforts of jamming aircraft, antiradiation missiles, and deception to attack the enemy's total capacity to defend from air attack.

4. Estimates for the support aircraft/strike aircraft ratio between the Air Force and Navy vary based on the war situation, availability of assets, and specific combat objectives. Generally speaking, Air Force ratios during Desert Storm were quite high (because of tanker, airlift, and electronic combat support) compared to Libya. Navy ratios, however, are constrained by what is aboard the carrier and the requirement to protect the fleet from air attack. Additionally, it is difficult to determine ratios based on what constitutes a "combat" sortie and a "support" sortie.

5. RAdm Riley D. Mixson, "Where We Must Do Better," US Naval Institute *Proceedings*, August 1991, 38–39.

6. The concept is discussed in numerous Marine and Army publications, and at least one joint Army/Air Force document: TRADOC pamphlet 525-5, *AirLand Operations: A Concept for the Evolution of AirLand Battle for the Strategic Army of the 1990s and Beyond* (Fort Monroe, Va.: Army Training and Doctrine Command, 1 Aug 91). This pamphlet is signed by the Army Training and Doctrine commander and Tactical Air Command's commander. Articles in *Military Review* (Maj Gen Stephen Silvasy, Jr., USA, "AirLand

Battle Future: The Tactical Battlefield," February 1991, 1–12; and Lt Gen Frederic J. Brown, USA, Retired, "AirLand Battle Future: The Other Side of the Coin," February 1991, 13–29) point to the evolution of Army doctrine to encompass the nonlinear, extended battlefield that is now reachable by Army air weapons as well as Air Force weapons.

7. Interview with an electronic combat laboratory engineer, Wright-Patterson Air Force Base, Ohio, January 1988.

8. There is considerable emphasis on Air Force doctrine at present among the academic community and upper echelons of command. There is much less discussion at the tactical level where it needs to be if doctrine is to have significant impact on a day-to-day basis.

9. During a roundtable discussion on the changes in tactical air power, with members of the CADRE Airpower Research Institute, an Air Force general discussed the differences in the use of sufficient, decisive, and overwhelming force as a result of Desert Storm. He indicated that, oddly, the success of Desert Storm may have done air power a disservice in future conflicts. The use of overwhelming force may not be physically possible in the future, based on projected drawdowns of US military strength, even though the American public will expect it.

10. Several papers have been written on the subject, one of which was presented to the TAF Electronic Combat Symposium held at Eglin AFB, Florida, in 1987. It is a perennial issue among the services, but while there is agreement that Air Force definitions are functionally more appropriate, there has been no change in the structure of the joint staff which plans—or would execute—JSEAD operations.

11. The primary joint memorandums that govern C^3CM and electronic warfare (Memorandum of policy nos. 185 and 95 respectively, both Secret) establish relatively firm relationships. It takes one to two years to coordinate among the various services an acceptable position that all will "sign up" to. This means that, even though the document may carry a very current date, it may be as much as two years old in thought process. This long cycle militates against a structure that is flexible and responsive to rapidly changing conditions.

12. *Tactical Air Forces Guide for Integrated Electronic Combat* (S) (Eglin AFB, Fla.: October 1987), 3-1-3-3. (Secret) Only unclassified paragraphs are cited.

13. The acquisition of the launcher avionics package (LAP), which is a computer-enhanced LAU-118 missile launch rail for the AGM-88 HARM, allows the F-16 to fire the HARM. On earlier F-16 models, the LAP computer "tricks" the F-16 into thinking the HARM is a Maverick AGM-65 missile and thus can use the Maverick firing logic and circuits. The more recent versions of HARMed F-16s (block 50-plus) have a more elaborate LAP program which interfaces directly with the F-16 fire director system and significantly improves the launch parameters. It is important to note that the F-16 HARM capability is not comparable in any way with the very elaborate and formidable capability represented by the F-4G. Current F-16 capability is more comparable to the Navy F/A-18 or A-6E HARM-firing capability. There are programs in motion to give the F-16 a more F-4G-like capability, but these are several years from fielding.

14. *Jane's Avionics, 1986–87* (London: Jane's Publishing Company, 1987), 149, and Martin Streetly, *World Electronic Warfare Aircraft* (London: Jane's Publishing Company, 1983), 25–26.

15. While undeniably better suited for the SEAD environment than any other single aircraft in the world, the F-4G is also a very specific, single-mission weapon system. The airframe which houses the Wild Weasel capability is a modified F-4E that has been fitted with antennas and very sensitive electronic detecting equipment as well as airframe modifications that allow it to fire the AGM-88. Considering that it is very expensive to operate and performs only a single mission, the F-4G appears less and less likely to survive in a fiscal and strategic environment that stresses economy and general capability.

16. Bert Kinzey, *The Fury of Desert Storm: The Air Campaign* (Blue Ridge Summit, Pa.: TAB Books, 1991), 50–51.

17. Norman Friedman, *Desert Victory: The War for Kuwait* (Annapolis, Md.: Naval Institute Press, 1991), 149–52.

18. "Iraq Attacks SEAD Plan Briefing" (S), Hickam AFB, Hawaii, August 1990. (Secret) Excerpted from unclassified portion of briefing.

19. Maj James L. Hendrickson, USAF, *Joint U.S. Army/Air Force Planning and Employment of Electronic Countermeasures* (Fort Leavenworth, Kans.: US Army Command and Staff College, 1978), 21.

20. Lt Gen Thomas W. Kelley, USA, Retired, Distinguished Lecture Series, speech, Air War College, Maxwell AFB, Ala., 5 September 1991.

21. The fire support coordination line, or FSCL, is the boundary to which army artillery fire can reach into the battle area. The Air Force has responsibility for the battlefield CAS beyond the FSCL. With artillery of longer and longer range and helicopter-delivered firepower projected beyond these traditional lines, the division of responsibility among Air Force, Army, and Marine SEAD is becoming extremely difficult to determine.

22. FM 100-5, *Operations* (Washington, D.C.: Department of the Army, 1986), 1.

23. TRADOC pamphlet 525–5, 15.

24. Maj Kenneth L. Travis, USA, *The Integration of US Army Electronic Warfare Capabilities in JSEAD Operations* (Maxwell AFB, Ala.: Air Command and Staff College, 1988), 19.

25. The AirLand Battle-future chart is presented in a Marine Corps publication, which further shows the evolutionary trend toward tactical and doctrinal convergence among the services in many areas. See Col Kent O. Steen, "AirLand Battle Future," *Marine Corps Gazette*, March 1991, 45–46.

26. This is another indication that the distinction between Army and Air Force SEAD roles is blurring as the future battlefield becomes more and more nonlinear.

27. TRADOC, US Army briefing, "Decade of Change—Challenge for the 1990s," Headquarters, Tactical Air Command, Langley AFB, Va., October 1991.

28. The Marine Corps is not the only service to exhibit parochialism as far as SEAD tactics are concerned. Parochial Marine viewpoints with regard to SEAD highlight the irony between a Marine philosophy that has traditionally melded land, sea, and air operations thoroughly and a Marine

viewpoint which has, until recently, steadfastly resisted using the same concepts with sister services.

29. The new Marine FMFM 1- series shows a peculiar mixture of pure Clausewitzian war strategy doctrine (FMFM 1, *Warfighting*) and classic Marine "nuts-and-bolts" combined-operations tactics (FMFM 1-1, *Campaigning*). FMFM 1-2, *The Role of the Marine Corps in the National Defense*, also does not define many other features of joint warfare such as C^3CM, EW, or joint air attack team (JAAT). The Marine concepts are integrated and keyed to maneuver (ground) warfare and, as inferred from the omission of SEAD as a "shaping" element in the battle, still are conceptually two-dimensional.

30. FMFM 1-1, *Campaigning* (Washington, D.C.: Headquarters US Marine Corps, 1990), 4.

31. Speech at Maxwell AFB, Ala., 17 January 1992. The Air Force's Air University subscribes to a nonattribution policy which precludes identifying the speaker by name for certain presentations.

32. FMFM 1-2, *The Role of the Marine Corps in the National Defense* (Washington, D.C.: Headquarters US Marine Corps, 1991), 1-1.

33. Steen, "AirLand Battle Future." Doctrinal issues currently proliferate all four services' professional journals, presumably in response to the radical changes in the fiscal and threat environment. It is clear that all services are moving toward each other in general warfare strategy and all use similar capability assets.

34. Speech at Maxwell AFB, Ala., Winter 1992. Nonattribution policy precludes identification for this particular presentation.

35. FMFM 1-2, 1-1.

36. Ibid., 6-2.

Chapter 4

The Merits of JSEAD:
The Quest to Achieve Effectiveness

Previous chapters have established five main points. One, SEAD's growing importance in air warfare is the product of a natural evolutionary process. Two, this importance is likely to grow as air power increasingly becomes a key element of national military strategy. Three, SEAD will grow in complexity as technology improves, and, paradoxically, will diminish as a specialized task because the costs of improved technology may prove prohibitive. Four, traditional measures of SEAD's effectiveness do not adequately address the reality of the changing world. Five, since we will need SEAD, we need to apply relevant criteria to determine which courses of action will yield the most effective results. Hence, criteria developed for SEAD must relate effectiveness to clearly defined objectives to be meaningful.

This chapter examines the personality of JSEAD by assessing some of its distinguishing features and determines the impact of individual military service cultures by using the models presented in chapter 2. The relationships among national strategy, theater strategy, operational art, tactics, and doctrine also play a key role in our quest for an adequate context in which to frame JSEAD.

The Personality of JSEAD and the Threat

Where does JSEAD fit, and why is a discussion of its personality important? JSEAD has many factors which distinguish it from other war-fighting skills.

First, JSEAD reflects a war-fighting mentality. It is more suited as a concept to fight wars than as a concept to maintain peace. As a concept and a tool, JSEAD exists primarily for the purpose of defending against or defeating an enemy in battle.

As such, the linkages between military strategy and tactical application are often more developed in JSEAD than in many other aspects of war fighting. JSEAD practitioners and staffs tend not to experience the steep relearning curve suffered by many other combat skills in the transition from peace to war.

Second, JSEAD is technical and highly specialized and focuses on specific, quantifiable, measurable phenomena. Because the language of JSEAD is electromagnetic physics, JSEAD tactics appear shrouded in mathematical equations and arcane phrases. This "personality trait" of the JSEAD community is a major contributing factor in isolating it from the mainstream of tactical thought where it needs to be.

Third, JSEAD planners' insistence on precision in definitions and objectives also has contributed—both negatively and positively—to JSEAD's personality profile. What appears to the general strategist as a precise objective statement often does not satisfy the requirements of a JSEAD planner. For example, a general objective of a strategic planner may be to "destroy the IADS structure of the Iraqi military." This is as precise an objective as the JSEAD planner is likely to get. The JSEAD planner's immediate concern will be the nature of the threat and how to neutralize it. This concern will most likely be framed in questions like: In what order do you want the enemy's IADS destroyed? How many resources will you allocate for this purpose? What is your eventual goal with respect to JSEAD? Will all services be included in the campaign, and, if so, what relationships do you want established? This is a level of detail the general strategist is often unable to deal with.

The decision makers in Desert Storm—unlike SEAD applications in Vietnam and Libya—answered these questions, and, largely because these questions were answered with the precision requested, SEAD planners were able to deliver a devastating SEAD campaign on the Iraqi IADS. The success of the Desert Storm JSEAD campaign demonstrates the successful linkages between strategic concept and tactical application forged as a result of forcing sufficient precision into tactical objectives. These procedures worked for Desert Storm, but the process was not institutionalized to codify these successful outcomes.

A deeper look at this phenomenon, however, reveals that it is unlikely that JSEAD's quest for precision will ever be codified. JSEAD planners' insistence on timely data and precisely defined objectives—because of the variety of factors and dynamics involved—may have created a requirement for a permanent ad hoc JSEAD planning structure. The dynamics of change militate against the creation of any structure that may hinder responsiveness to the rapid tempo of modern combat.

JSEAD is fundamentally pragmatic and realistic in its approach to the IADS threat; it is concerned with the survival of friendly air power assets while neutralizing the enemy's ability to defend from air attack at the most basic level. The JSEAD planner focuses on what enemy weaponry can and cannot do and what will and will not "work" against the enemy's IADS array. The JSEAD practitioner is trained to make a realistic appraisal of the enemy's ability to deny air superiority to friendly aircraft. This appraisal must be devoid of bravado, timidity, or exaggeration. As a pragmatist, the JSEAD planner's first course of action is to locate, define, and analyze the threat in terms that make sense to tactical planners. In this process, JSEAD planners attempt to bond strategy to tactics in a way which allows decision makers to determine how to wage the air campaign most effectively.

Nowhere is the difference between JSEAD *strategist* and JSEAD *tactician* shown more clearly than in the initial questions they ask of the war-fighting situation. The first question the JSEAD *strategist* asks is "What is the *objective*?" The first question the JSEAD *tactician* asks is "What is the *threat*?"

Military lexicon has elevated some phrases to almost metaphysical levels. "The mission" is such a phrase for tactical aviators; "the threat" is such a phrase for JSEAD planners and the electronic combat community. The threat's characteristics, general nature, inner workings, technical aspects, doctrinal use by the enemy, and role in the enemy's intention to wage war and defend attacks are at the heart of the JSEAD planner's concerns.

The evolutionary nature of the threat, described in chapter 1, illustrated that the threat changes in identifiable ways over time and exhibits a sense of direction. However, the IADS threat's evolutionary process is not linear; *evolution*, as a defining term, belies its dynamic essence. While each particular situation is different, the general evolution of defensive capability against air attack has accelerated so rapidly that the term "revolution" may be more appropriate. Figure 16 depicts linear versus nonlinear evolution and shows the problems revolutionary change poses for the tactician preparing for the threat.

Figure 16 is not intended to give quantitative measurement but to show the effect revolutionary change has on one's ability to predict and prepare for the future. As one can see from the graphs, revolutionary change produces a much larger area of unpredictability than evolutionary change. This unpredictability makes classical methods of developing tactics and theater war strategy largely suspect because there is increasingly no definitive baseline against which to plan. We need a more relevant way to measure desired effects, and, in light of an increased inability to predict future needs, we may have to reconsider acquisition strategies directed at countering anticipated capabilities.

The future threat will be highly integrated. Strategy and tactics analysts will find it increasingly difficult to distinguish where air defense ends and command, control, and communications begins. This has enormous implications for developing target lists and planning strategy. In previous wars, critical node analysis was a straightforward process of identifying targets to bomb, aircraft to destroy, bridges to blowup, and SAM and radar sites to neutralize, among other things. The emphasis was placed on physical identification and destruction of specific sites, not on a systematic analysis of the overall enemy air defense structure.

As the future threat becomes less a "place" and more a collection of possible scenarios, total IADS degradation will become very important as a principal goal. A SEAD campaign's objectives, therefore, must consider many more courses of action because of the expanding nature of "the turf." Destruction will need to be considered as part of an overall plan to degrade the enemy IADS rather than an end in itself.

The future air defense threat will have at its disposal the most sophisticated technology available in the world. The proliferation of high-technology air defense systems—

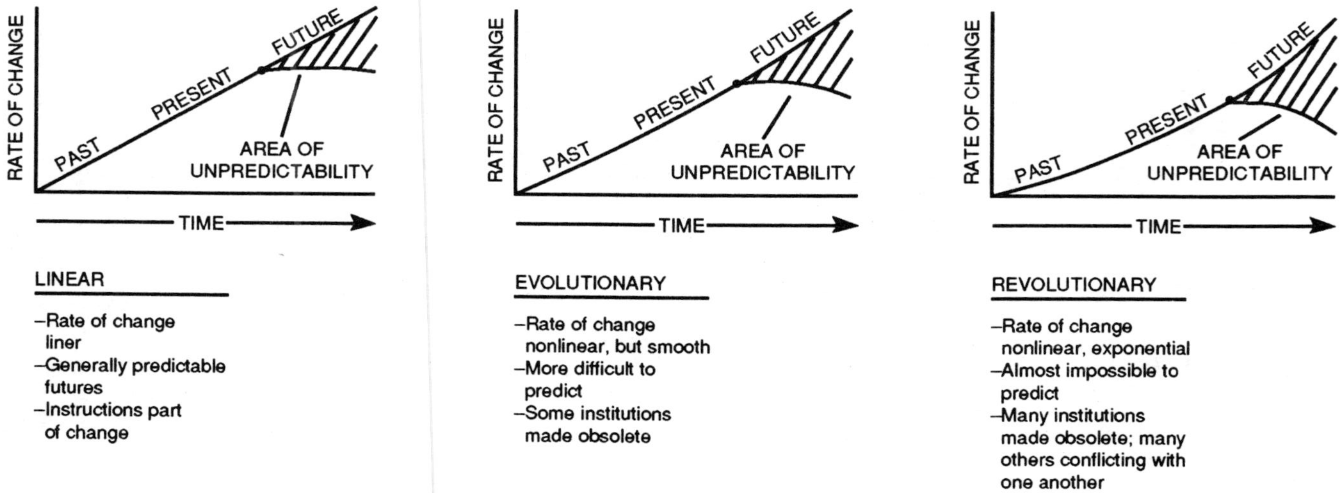

Figure 16. Linear versus nonlinear SEAD tactics evolution.

including entire networked arrays complete with trained technicians to operate them—is likely to increase for two related reasons: the demise of the Soviet empire (which has a need for hard currency and is willing to sell off much of its weaponry cheaply) and the entrepreneurial impulses of Western arms producers. Together these have produced an explosive situation where virtually any nation wishing to arm itself with state-of-the-art weaponry may do so. Even lesser-developed nations of insufficient means and great internal instability—with the acquisition of an inexpensive IADS and rudimentary ballistic missiles of moderate range—can pose credible threats to anyone anywhere.

The days of the totally "Red" threat (systems of Soviet-Warsaw Pact manufacture—from telephone to tactical missile) are gone. In their place are potential enemy IADS systems that are a highly technical hybrid composed of Soviet (now fragmented into nearly a dozen arms-selling entities), French, British, German, Italian, Japanese, and even "frontline" American systems.

An analysis of currently existing weapons systems, radars, communication nets, and computers sold to "potential adversaries" shows literally thousands of different systems scattered across the planet in unique configurations that change on a daily basis.[1] For instance, the communication net built by "friendly powers" in Panama prior to Operation Just Cause was a hybrid system of great complexity and technological sophistication. The net posed enormous tactical problems for the electronic combat community, most of whose plans dealt with systems built by Warsaw Pact manufacturers.[2]

The future threat is making a subtle shift from three to four dimensions in its application to war. This shift has ominous implications for air power and theories of maintaining air superiority over the battle area. Time (speed, duration of war effort, pace, etc.) has qualitatively changed the character of war. The speed of war as driven by computers, detection and tracking systems that relay data at the speed of light, hypersonic ballistic missiles, and large numbers of supersonic weapons systems capable of pinpoint accuracy has introduced time itself as a critical dimension of the future threat. In the continuing contest between offensive air power and defensive

IADS structures to gain ascendancy—one over the other—it appears the defensive is asserting itself. The future IADS may soon forge ahead of air power in its ability to defend successfully against air attack largely as a result of its success in making the transition to a true four-dimensional war-fighting doctrine. The technological chess game between air superiority and air deniability continues unabated. Strategists must approach the sophisticated IADS of the future respectfully if offensive air power projection capability is to retain its military worth.

According to the defensive/offensive matrix in chapter 3, as it evolves, JSEAD appears more effective as an offensive tool. However, war planners and practitioners alike traditionally have viewed SEAD as a defensive mechanism to defend aircraft conducting offensive operations. While this defensive application of SEAD has been the norm, the increasingly technical and integrated nature of the potential enemy's air defense structure and command-and-control apparatus mandate a more encompassing and proactive role for JSEAD in the future if the air offensive is to maintain credibility.

The proactive JSEAD role envisages a transition from "defensive-offense" to more "offensive-offense" applications of JSEAD. In this capacity, JSEAD's aim is to strip away the totality of an enemy's ability to defend from a generalized aerial attack. The action is designed as an active measure to force the enemy into predictable modes of behavior. JSEAD takes the initiative away from the enemy and obligates manageable behavior by limiting the enemy's range of options. While recognizing that SEAD will always retain a defensive role, we must also consider the potential use of JSEAD as a primary element of offensive strategy.

The World of "Joint" and JSEAD

Retired Air Force general Michael Dugan pointed out in a magazine article that, while each service fills specific niches associated with its fundamental role, the requirement for joint cooperation is compelling.

> Modern warfare is joint warfare. The US Army and US Air Force have fully committed themselves to orchestrating and synchronizing operations—along with those of the US Navy and Marines—to fight the theater commander's war. At the same time, all recognize that aerial, naval and land warfare are fundamentally and necessarily different. To maximize the contributions of each, the key commanders must exploit the services' different capabilities. In the profession of arms, teamwork is fundamental to success. It results not only from a thorough knowledge of our own jobs, but from understanding the strengths of all and where each fits to best contribute to victory.[3]

The analysis of how each service perceives and uses SEAD in the execution of its strategy and tactics (presented in chapter 3) serves both as a test of SEAD's effectiveness and as a gauge of whether the overall needs of the theater commander are met. Furthermore, the evolving nature of war demands a more authentic joint approach in the application of SEAD than has been the case in the past. Even though practitioners of SEAD at the tactical level have been "doing it jointly" for at least a decade, individual service doctrine has lagged this effort considerably.

Many military observers of Desert Storm seemed puzzled at the rapid adaptability JSEAD displayed during war operations. JSEAD did not suffer the steep learning curve many other elements of the joint staff experienced. While Desert Storm was a "success story" for jointness, that JSEAD's rapid adaptability came as a surprise is revealing.[4] An obvious information shortfall existed between those who formulated general theater strategy and those who executed specific SEAD tactics. This paradox seems to support the concept that tactics execution on the battlefield and strategy formulation at headquarters tend to operate in cycles isolated from one another. Put another way, the fact that strategists were surprised by the ease with which electronic combat-supported SEAD adapted to joint operations supports the oft-quoted proposition that the study of American fighting doctrine by a potential enemy is useless because American warriors will abandon doctrine and strategy if they assess it not to be working. *Strategy and doctrine are conceptual; operational art and tactics are pragmatic.* Nonetheless, if a criterion of effectiveness linking SEAD strategy to tactical objectives is to have combat merit, doctrine and strategy need to be connected to theater operational art and tactics in a meaningful way to make true JSEAD a reality.

From a functional perspective (i.e., waging war effectively), it does not matter where the impetus to integrate strategy, doctrine, operational art, and tactics comes from with respect to JSEAD. The impetus can stem from directives generated by the command structure from the top down or mandated by tactical success (or failure) from the bottom up. A strong argument can be made that the apparent success of SEAD tactics to influence overall strategy from "the bottom up" indicates that tactical imperatives—not doctrine—were the real drivers during Desert Storm. If substantiated, such a revelation would raise serious questions about the relevance of traditional models used to explain the flow of command and the SEAD decision-making process in the post-cold war era.

The cybernetic model introduced in chapter 2, where each element of the decision-making mechanism influences the others in achieving balance, explains more readily why the process appears to be influenced from the bottom up. Figure 17 depicts the process.

A comparison of the traditional vertical model and the cybernetic model describing the SEAD process shows several differences which may account for JSEAD's surprisingly rapid but effective transition from peace to war. First, the traditional vertical model is an idealized concept. It shows how a perfect system, with perfect leaders, perfect information, and a perfect data-transfer system from one level in the decision-making process to the next should operate. The "real" JSEAD process is far from this idealized concept. Because it is imperfect, the JSEAD process requires a feedback mechanism to assess whether it is accomplishing the task decision makers set before it. Defense suppression and the JSEAD process that evolved in Desert Storm were based on tactical pragmatism and drove strategy based on what *really* worked as opposed to an idealized concept of what decision makers *desired*.

The role feedback plays is significantly different in the cybernetic model. The traditional vertical model allows for feedback only where the edges of each layer of the decision-making process meet. The cybernetic process is interactive

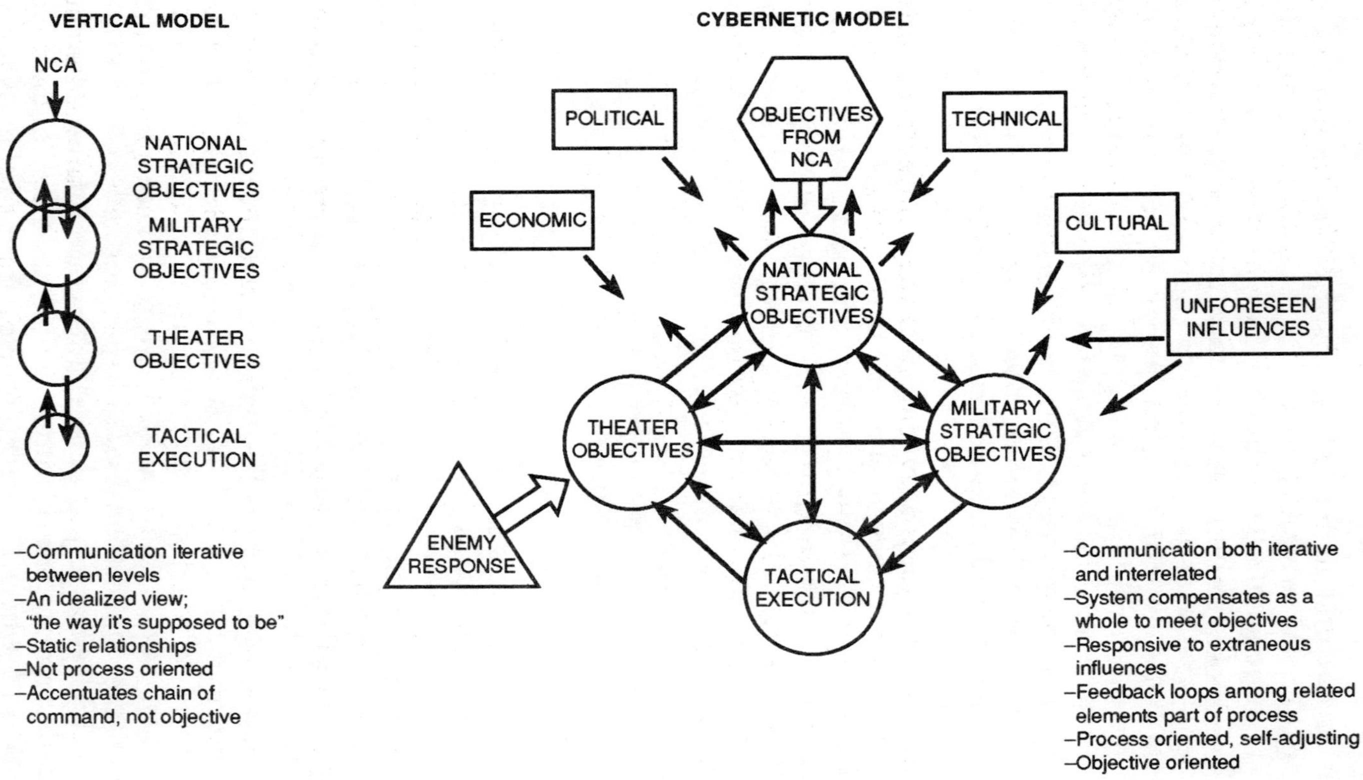

Figure 17. Differences between vertical and cybernetic models.

among each element depending on whether the process is meeting the objective set before it. Feedback within the process itself adjusts the behavior and choices of the system to meet the overall objective. For instance, let us assume that 10 F-15Es were lost during the first night of the IADS campaign in Desert Storm. The loss of so many high-value aircraft would have had a *simultaneous effect on all aspects of the process and the entire system would have readjusted itself to solve the problem*—not just the bottom end of the vertical stack.

The JSEAD process, as it evolved in Desert Storm, was interactive, not vertical. The electronic combat planning cell was able to translate overall military objectives into meaningful SEAD tasks that, once accomplished, could be measured against the objective set by the joint force commander. A relevant criterion based on clearly defined objectives provided—in most cases—feedback that kept the JSEAD process focused on the real problems of defense suppression at the most basic levels of combat.

The cybernetic model presented in figure 17 provides a frame of reference to assess effectiveness and to define the evolving JSEAD process at work. The objective criteria at the macro level ask some penetrating questions about the nature of JSEAD decision making in the war-fighting context. A return to basics, using an adaptation of Philip Crowl's model,[5] will help us define what we need to evaluate JSEAD effectiveness in meaningful terms. Crowl stresses the primary dictums of classical military thought expressed clearly from Clausewitz to Norman Schwartzkopf: define the objective, know the objective, achieve the objective. This is a conceptually elegant, simple, and easily understood primary dictum.

Part of the difficulty in assessing how to evaluate JSEAD effectiveness lies in the disconnect between JSEAD's strategic purpose and its tactical application. Fundamentally, JSEAD application seems to be a tactics-driven activity, but the *reasons* for its application are strategic. Both strategic JSEAD and tactical JSEAD are influenced by doctrine, but in different and oddly contradictory ways. Joint doctrine, a concept generated to aid overall national military and theater strategy where more than one service is employed, pulls national and theater strategic concepts in one direction. Individual service doctrine, however, tends to pull theater operational art and tactics in another. The disconnect this creates is the result of two competing cycles' being polarized by doctrinal outlooks that work at odds with one another.

It is ironic that "joint" doctrine and "service" doctrine should operate in opposition in the war-fighting process, but they appear to do just that. The links are tenuous.

The disconnect between JSEAD strategy formulation and single-service-applied SEAD tactics is obscured if we use the vertical model to analyze the process but is clearly visible if we use the cybernetic model. Figure 18 shows the tensions created by joint strategy JSEAD doctrine and single-service tactical SEAD doctrine.

The tension between joint doctrine and single-service doctrine often pulls the SEAD decision-making process apart. The disconnect, for JSEAD, most often occurs when high-level decision makers attempt to translate national military strategy into theater strategy and operational art. The differences between the joint perspective and individual services' perspectives make this translation very difficult. The result is that JSEAD "results" often tend to be ambiguous; the linkage between tactical effects and achieving overall objectives is often obscured by the transition from jointly formulated *strategies* to single-service applied *tactics*.

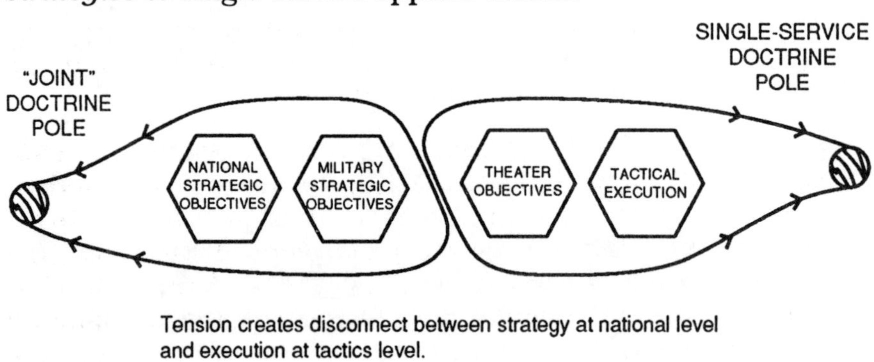

Figure 18. The tension between joint and single-service SEAD.

Brave New World: "True" Joint SEAD

In the transition from a military force built on the collective capabilities of each of the separate services to one which will likely require a pooling of resources, the basic definition of "joint" appears to be changing. The definitions currently used range from "the application of force to meet specific objectives which require the assets of more than one service" to "an encompassing concept embracing holistic patterns of systems and tactics application in the prosecution of overall objectives which may employ the assets of one or more of the respective services." The requirement for a definition of "true" JSEAD is critical for setting the JSEAD context of the future world.

Perhaps the best way to arrive at a definition of "joint" in modern warfare, for JSEAD purposes, is to describe some of the things it is not. "Joint" is not merely the *sum* of all weapons, tactics, strategies, doctrines, histories, traditions, and parochialisms of the respective services. It is not the total domain of a joint, unified commander and staff. It is not a series of iterative SEAD strikes using "the right tools in the right way to achieve the objective," though this comes close.

Each of the services contributes—by its mission and roles—to the joint menagerie of tools and tactics, but the process which melds them together is more than the sum of their parts. Jointness is an evolving paradigm. It is more a *working concept* than a "nuts and bolts" allocation of weapons to target sets, more a *dynamic process* which relates specific means to general ends than a checklist of tasks to be performed by various joint battle staff agencies.

General Dugan's statement that "modern warfare is joint warfare"[6] compels us to seek the characteristics in the modern realm that separate us from previous times. Succinctly, these are limited resources, growing technological diversity and complexity among friends and enemies alike, unstable economies, regional instabilities, and an American military that is shrinking to an indeterminate size because of the demise of the Soviet Union and the accompanying perception that the threat has diminished.

The imperative to wage war jointly is not just doctrinally attractive and politically palatable; it is one of the only effective ways left to the US military to fight wars of the future. Because air power will become a critical factor in future US military power projection, JSEAD becomes all the more crucial. Therefore, an understanding of jointness as it applies to SEAD takes on a special meaning for future American power-projection capability.

Joint Publication 1, *Joint Warfare of the US Armed Forces*, which defines the general view of joint doctrine and how it relates to war fighting, is quite clear in that principles of war—including those to which SEAD contributes—are decisive only as a result of unified, coordinated effort. The publication goes on to state:

> The principles of war represent the best efforts of military thinkers to identify those aspects of warfare that are universally true and relevant. The principles of war currently adopted by the US Armed Forces are *objective, offensive, mass, economy of force, maneuver, unity of command, security, surprise*, and *simplicity*. These principles deserve careful study by all who practice the military art, because the insights suggested by their analysis span the entire range of military operations. . . . In all cases, the principles are applied broadly, avoiding literal or dogmatic construction, and with due regard for the unique characteristics of joint warfare.
>
> The first application is **unity of effort.** Success in war demands that all effort be directed toward the achievement of common aims.[7]

JSEAD, as a process of convergent evolution among the services, reflects the essence of unity at the tactical level. However, the concept cannot be applied until each level of the decision-making process—from national command authorities (NCA) through the joint chiefs to lowest tactical levels—can translate its objectives into operational terms usable by the next successive level in the process.

The difficulty for JSEAD planners and tacticians is bridging the gap between strategic concepts and tactical execution. The criteria of success which enable the NCA to determine if they have met their objectives and the measurements of effectiveness that an electronic combat cell of a war-fighting commander's staff employs to determine whether their JSEAD campaign was successful look much different. Yet they must relate directly to each other if the "unity-of-effort" principle is to obtain. Note that the objective of jointness is, through unity of effort, to achieve an objective. The imperative is not *who* commands but *how* we use our resources *together* to meet the objective. This is why a measurement of effectiveness that encompasses the entire process is required.

Effective JSEAD must, first and foremost, translate overall objectives into usable terms that employ the collective assets and tactics of all the services. There must be a single point of contact and a single command structure that directs the efforts of pooled resources.

The scale of measurement JSEAD planners use must satisfy the overall objectives of the battlefield commander. The scale must be able to answer the question "What is the most effective way of employing JSEAD to protect air power projection and neutralize the enemy's capability to defend from air attack?" The measurement scale also must be able to determine, after the fact, whether a particular weapon or tactic "worked." Additionally, the scale must be able to determine not only if the SEAD weapon, tactic, or strategy was effective but *how* effective a particular employment option was.

The evolving nature of JSEAD indicates that the entire framework we have used to evaluate JSEAD's military worth may have outlived its usefulness. A new JSEAD criterion that relates to "the world of jointness" is emerging, and it appears more suited to cybernetic process analysis than to traditional vertical "chain-of-command" models.

The Four Continuums: How to Use Them

The determination of criteria in the JSEAD process can best be approached by analyzing the four continuums outlined in chapter 2 and placing each of the services within the context to determine effective employment options. These continuums were the "need-/resource-based" continuum, the "threat-/capability-based" continuum, the "piecemeal/integrated tactics" continuum, and the "defensive/offensive" continuum. Each of the services fits somewhere in each of the continuums, and there are strengths and weaknesses associated with each service's SEAD approach. Each continuum is related to the others, but the relationships are characterized by variables which change in nonquantifiable and often unforeseen ways. Nonetheless, we can make some guarded assessments about how these general relationships alter the measures of effectiveness. By analyzing what is needed to meet the specific SEAD objectives for the specific situation, one can gain a clearer idea of what service's (or combination of services') assets, techniques, procedures, and tactics would best suit the war-fighting theater's purposes.

The Need-/Resource-based Continuum

The need-/resource-based continuum shows a wide variation among the services. The Marines and Navy are more accustomed to making do with what they have (Given what I have, what can I do?). This leads to more innovative approaches to SEAD, but the tactics/weapons mixes may tend to cause higher attrition or narrow the range of capability against the IADS. Army doctrine commits them to a course of acquisition based on perceived needs (Given what I have to do, what do I need?). Nevertheless, the Army is practiced in the art of executing tactics and strategies based solely on what they possess. The fact that Army leaders are familiar and comfortable with their doctrine enables them to use what they have, though doing so may limit their tactical flexibility. Air Force SEAD practitioners, being more technologically oriented, tend to view the future in terms of what assets they need to accomplish the goal. While this approach provides the Air Force with superb SEAD equipment, it often places its leaders in the unenviable situation of fighting today's wars with tomorrow's weapons. Additionally, with the shrinking military budget, the mandate to purchase sophisticated hardware based on a visionary doctrine will become more difficult to justify when older or existing technologies appear sufficient. Nonetheless, it is clear that advanced technology is required for the application of decisive force in the power-projection role as enemy air defense represents one area that, ironically, will become *more* technologically complex with the demise of the Soviet empire.

Figure 19 depicts the need-/resource-based continuum. The scale is not intended to reflect a quantitative measure; it is intended to provide a basis of comparison of each service. The horizontal axis displays the relative range between resource-based and need-based approaches to the SEAD problem. The vertical axis shows the relative effect that limited resources have on innovation with respect to SEAD weaponry and tactics.

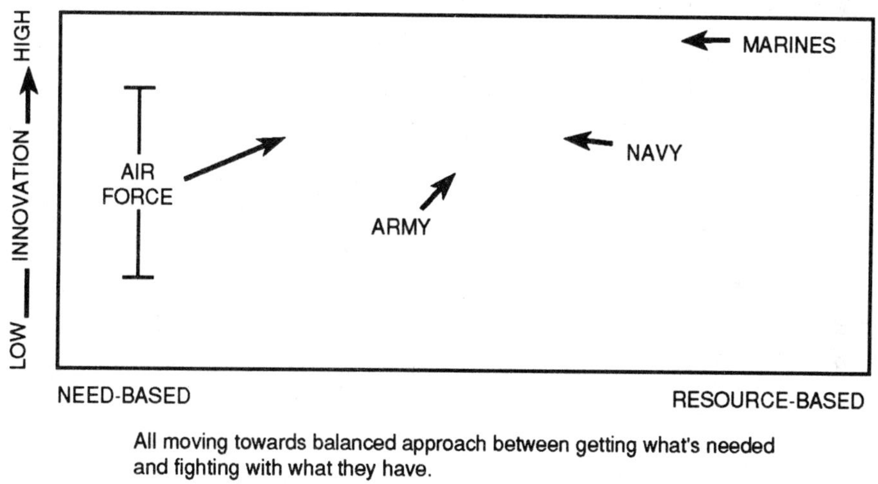

Figure 19. Need-/resource-based continuum.

Two conclusions can be drawn from figure 19. First, a need-based approach (represented by the traditional Air Force position) tends to exist only in a relatively resource-rich environment. Second, limited resources force a more creative approach to solving SEAD problems. The corollary to these observations is the tendency for resource-rich, need-based approaches to SEAD to be relatively less innovative than resource-poor, resource-based approaches.

There is a notable exception to the creativity-resource relationship noted above. The Air Force's impetus to integrate electronic combat and SEAD assets was—and is—an innovative

SEAD concept that presages JSEAD. Ironically, it was spurred by factors that have little to do with this continuum. Because the Air Force had developed so many separate EC and SEAD systems (the result of resource-rich, need-based concepts), the EC community was forced to develop a scheme to integrate them to make maximum use of their expenditures.

Political and economic pressures promise to force JSEAD toward the resource end of the continuum. The Navy philosophy of "multiroling" aircraft for SEAD functions is being adopted by the Air Force and Army. The next step in the process will occur when traditional single-service SEAD assets can no longer provide even minimum protection for their forces given resource constraints. This will compel the military to move toward a "truer" JSEAD—not because politics (the Goldwater-Nichols initiative) mandate it but because necessity will demand it.

A key question for JSEAD practitioners related to the push toward the resource end of the continuum concerns the absolute limits of innovation to compensate for constrained resources. For JSEAD, "doing more with less" has absolute limits. The difficulty lies in defining those limits, and once they are defined, convincing decision makers that they are real. Unfortunately, the only way to verify that resource-based JSEAD tactics and assets might not be effective is after the fact, after friendly forces suffer attrition rates that blunt the effectiveness of air power as a power-projection tool in a shooting war.

Another important related issue in the analysis of the resource continuum is the question of goals in power projection: Do decision-makers choose JSEAD forces that are merely sufficient to meet objectives, or do they construct forces which reflect decisive—even overwhelming—superiority over the potential enemy? More importantly, how does the decision maker determine what "sufficient," "decisive," or "overwhelming" mean in operative terms with respect to JSEAD? Desert Storm's success seems to indicate a strong preference for building and using overwhelming force in the prosecution of war. In this context, the decision maker either composes forces that are "overwhelming" by number and firepower or develops strategies and tactics that use available resources in an attempt to overwhelm the enemy. This latter case for air power (and JSEAD by implication) was the thesis of an *Aviation Week and Space Technology* article.[8] More fundamental still is the issue that surfaces when, having reached the limits of innovation in the resource-constrained environment, the US military determines that it is no longer able to guarantee even sufficient JSEAD for air superiority. When faced with the political decision to commit American military power in these situations, how will military leaders respond?

In order to avoid crossing the line from sufficient to insufficient JSEAD capability, JSEAD strategists must clearly define what constitutes the absolute limits beyond which—no matter how we reorganize, innovate, or reconstitute pooled JSEAD assets, tactics, and strategy—we are no longer effective in neutralizing the enemy IADS.

The Threat-/Capability-based Continuum

The threat-/capability-based continuum is closely related to the need-/resource-based continuum. In response to the perception of a diminished Soviet threat, the American political process has mandated significant reductions to the military. The same process that is responsible for pushing JSEAD toward the resource end of the need-/resource-based continuum is also responsible for pushing JSEAD toward a philosophy that emphasizes general capability over specific threat-based strategies. The JSEAD threat, unlike the uncertain world of the future, has identifiable characteristics. The requirement for a generalized capability is a military requirement against the relatively sophisticated JSEAD threat of the future. The diversity of technology available to regions where US forces will most likely be engaged will require electronic equipment, weaponry, and tactics that will be able to respond to a wide range of enemy capabilities. But, oddly, the impetus driving JSEAD towards capability-based concepts is the politico-economic environment.

The threat-/capability-based continuum displayed in figure 20 shows an interesting distribution. As a function of time, the services are moving toward each other—toward the

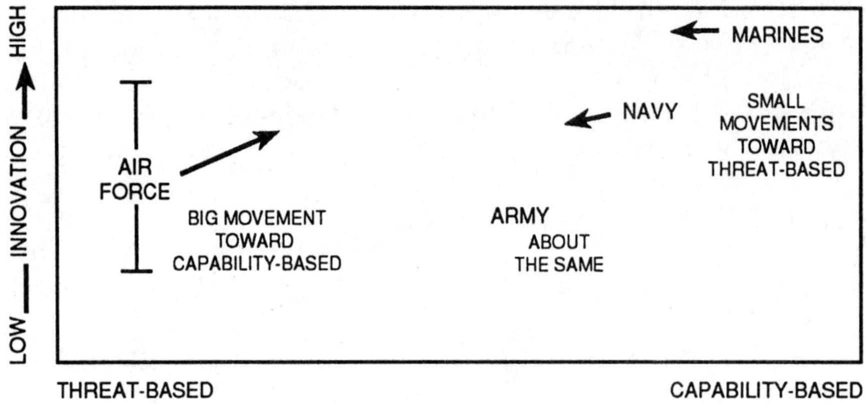

TRAITS

–Specific weapon/specific threat
–Many kinds of weapons
–Focus on one weapon/tactic
–Tactics codified
–Complex, dogmatic doctrine
–Specific outcome stressed

TRAITS

–Generalized forces/broad capabilities
–Fewer general types of weapons; versatility stressed
–Focus on mix of weapons/tactics to achieve objective
–Tactics as guidelines
–General, easily understood doctrine
–Overall objectives stressed

Figure 20. Threat-/capability-based continuum.

capability-based end of the spectrum. The Navy and Marines have always tended to this end. The Army has always tended toward the middle range and the Air Force toward the threat-based end of the spectrum. The Air Force is moving toward the capability-based end for two primary reasons: an uncertain threat and diminishing fiscal resources that militate against acquiring mission-specific weaponry. A generalized capability-based JSEAD—though explainable on the basis of economic pragmatism—is dangerous because it makes American air power vulnerable to technological surprise and increases the amount of assets (hence the potential for increased casualties and attrition) required to perform a given function. However, the move toward capability-based JSEAD provides opportunities in much the same way as the resource-based criteria does in that it forces a more flexible distribution of assets and it aids in the development of more innovative tactics.

The Piecemeal/Integrated Tactics Continuum

The piecemeal/integrated tactics continuum presented in figure 21 is complex. The Air Force clearly stands at the integrated end of the spectrum, and the Navy and Army tend toward the piecemeal end of the spectrum. The Marines fit, oddly, at both ends of the spectrum. They exhibit piecemeal tactics and weapons in their SEAD application because they tend to use capability-based tactics like the Army and Navy, but they also use a fully integrated tactics approach as a function of Marine combined-arms strategies. The strength for JSEAD is clearly in the integrated category, though piecemeal applications may suffice in situations where holistic approaches are not required (very limited threat array or very limited force objectives). Air Force approaches to JSEAD, though not completely consonant with jointness, do meet the criteria for the most effective consolidation of assets, if not tactics.

The Air Force lead in this area provides a natural rallying point for general JSEAD integration concepts. However, the Air Force's integrated-tactics concept will need to be modified by resource-based and capability-based concepts of the other services to make full use of this Air Force EC concept.

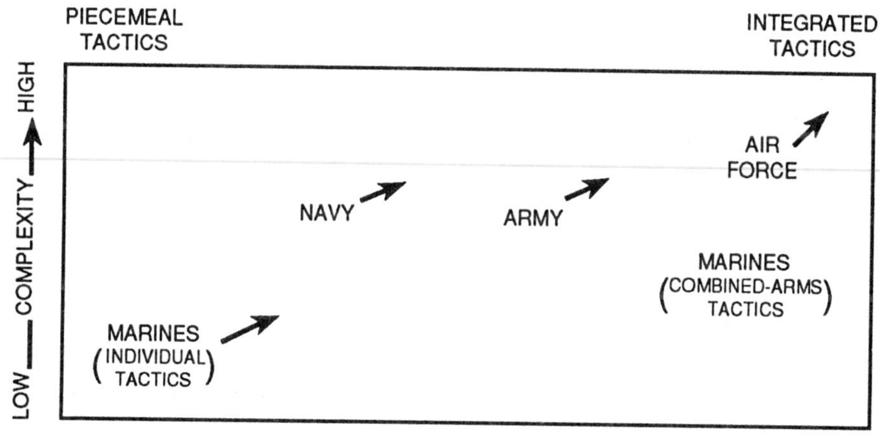

–Impetus to complex, integrated tactics
–Will create training problems
–Joint-/single-service tension in doctrine/application of forces will create problems

Figure 21. Piecemeal/integrated continuum.

The Defensive/Offensive Continuum

The defensive/offensive continuum poses some unique problems in applying measures of SEAD effectiveness. The Navy understandably represents the more defensive end of the spectrum, while the Air Force tends toward the offensive. The Army and Marines tend toward the center. This continuum reflects the doctrinal predisposition of each of the services. The effectiveness of the offensive or defensive predispositions, in real terms, depends on what the war-fighting situation calls for. When the NCA determines a defensive posture to be in our best interest, Navy-Marine conceptual frameworks would apply best. If overall objectives require the application of sudden, strategic offensive power, the Air Force concept of operations would be better suited for JSEAD.

Figure 22 is an adaptation of the Clausewitzian defense/offense matrix contained in chapter 2. The principal concept shown is that, through the application of JSEAD measures, all of the services' offensive capabilities are enhanced. This is because JSEAD's general evolutionary path has proceeded from a defensive to an offensive posture.

The four continuums, when taken together, enable us to build a more encompassing picture of JSEAD applications and potential effectiveness as a tool in war. The continuums set the context, give the tactician a way of placing JSEAD in perspective in a changing, turbulent time, and provide a means of comparison in bridging the gap between the old, familiar world and the uncertain one we live in.

The General Context: A New Way of Looking at JSEAD

Currently, JSEAD is described in three categories: campaign, localized, and complementary.[9] These are broad concepts that coincide with the planning process in the development of joint operations and contingency operations plans. JSEAD campaigns are related primarily to theater strategy which targets the IADS infrastructure of the enemy. Localized JSEAD is related to the tactical level of theater operations and is geographic and/or time specific. Complementary JSEAD is most often associated with "pop-up targets" (targets of opportunity) and the sudden need for self-defense from an unforeseen enemy air defense system. These macroconcepts are explained in the JSEAD *Multi-Service Procedures for the Joint Suppression of Enemy Air Defenses* manual published in 1990.[10] Theater strategic (campaign) JSEAD is generally the province of Air Force

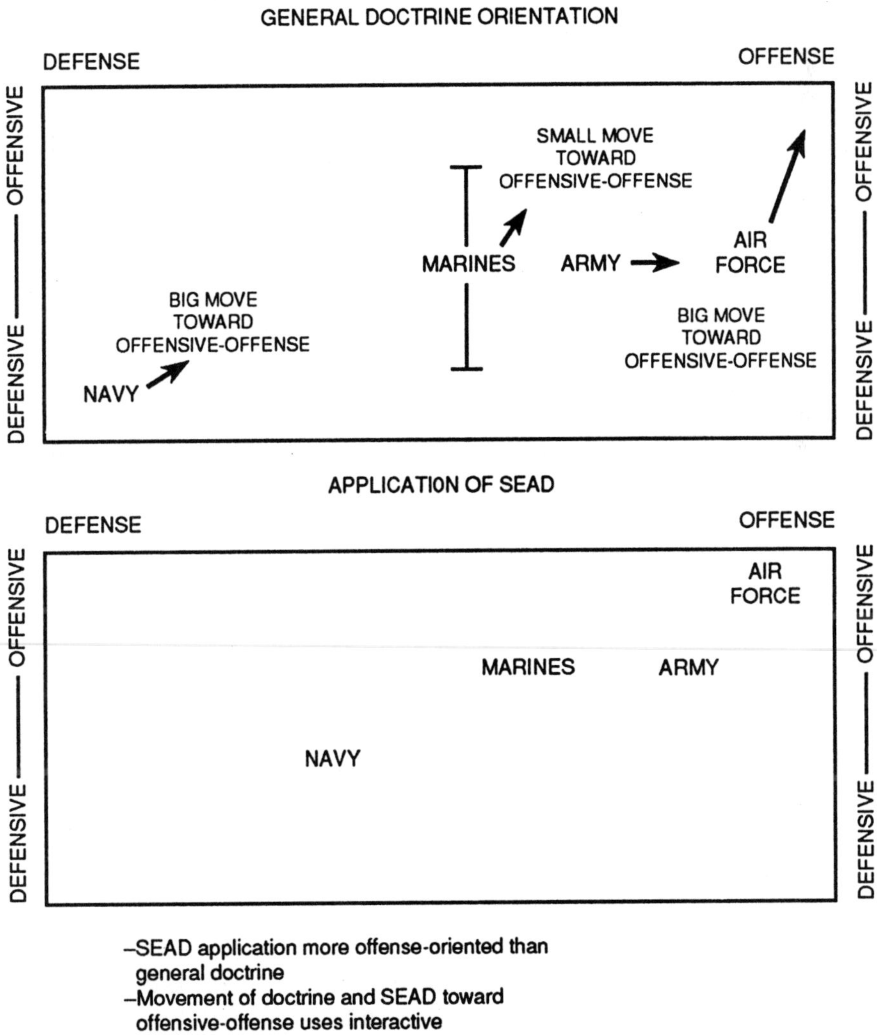

Figure 22. Defensive/offensive continuum.

thought, tactical JSEAD in localized areas is a practice of the Air Force and Army, and complementary JSEAD is used most commonly by Marine-Navy aviators in the power projection and fleet defense role.

Another way of viewing applications of JSEAD is by assessing what would be most effective in a high-, medium-, or low-intensity conflict (within the context of FM 100-5). Depending on the complexity, size, quality, and disposition of the enemy IADS and organic field army air defenses in each of these cases—and what the friendly theater commander specified as the key military objectives against the enemy—the meld of SEAD forces, sequencing, tactics, and weapons could vary considerably.

"Intensity of conflict" (another way of categorizing SEAD applications)—as clear a term as it appears—is complicated for JSEAD. The intensity of a war is related to tempo, size, violence (lethality), and politico-military ramifications for world/regional stability. The mixture of influences, only some of which the military has the power to control, defines not only how intense a war is but who controls the assessment of intensity. The level of effort that will be brought to bear during a war of a specific level of intensity is often beyond the purview of the military—at least during the initial stages of conflict. A "high-intensity war" is not necessarily "just a major war" requiring large numbers of forces. It could be a short war requiring a relatively small number of forces which expend ordnance with extreme violence.

For JSEAD, however, a high-intensity conflict invariably means the application of jointness in its prosecution. Medium-intensity conflict may require JSEAD or single-service applied SEAD, and low-intensity conflict may or may not require SEAD at all. It is not peculiar, therefore, that various squadron SEAD assets (especially among the Air Force and Navy) possess some of the most complex, incremental "packages" for contingency operation plans (CONPLANS) and operations plans (OPLANS) that exist.[11]

Both measurements—the determination of intensity and the characteristics of the threat—require a tool to determine how to allocate, organize, and direct SEAD assets to prosecute the theater commander's objectives. Making such an assessment also would require a framework in which to view the pertinent variables and criteria at work.

Using "high, medium, low" as a scale on one side and "complexity, size/quality, disposition" across the top, we form a three-by-three matrix. By using the matrix, one can establish a context from which to make a decision based on the criteria at work.

The matrix in figure 23 accommodates all the variables and criteria needed to identify the most effective tactics, strategy, and resource combination in a given situation. The terms require explanation to increase the matrix's usefulness.

> *Complexity:* The extent to which an enemy system is integrated, contains state-of-the-art systems, is trained according to a doctrine, and possesses a sustainable, reliable logistics apparatus to maintain the IADS.
>
> *Size/Quality:* The number of systems the enemy employs and the quality of the individual systems. Density of threat is also covered under this category. From a planning/fighting perspective, one must balance available resources against needed resources in order to achieve specified ends. Often times, tacticians discover that a straightforward comparison of an enemy's size/quality SEAD index to our own is invalid because the enemy may be vulnerable in areas that have nothing to do with quantitative/qualitative comparisons.
>
> *Disposition:* An estimate of the relative distribution of assets over space and the composition of the enemy's air defenses. An enemy IADS consists of discrete pieces of equipment located on specifically identifiable territory. Its distribution is affected by geographic size and type of terrain. EC specialists can ascertain a great deal about an enemy's thought process and IADS firing doctrine by analyzing this attribute. Disposition of enemy IADS also answers the more pragmatic targeting questions concerning critical node analysis, distances to targets, what to avoid, and what to suppress.

This matrix *provides a framework to assess the relationship between the objective and how it is to be met.* The matrix also attempts to consolidate graphically the concepts which account for the impacts of service culture, doctrinal differences generated by role, and the requirements of the theater commander to meet war objectives by using combined resources.

It is entirely possible for the war situation to contain mixed elements within the matrix: a high-complexity threat array does not necessarily also have to be a high-size/quality or

INTENSITY OF CONFLICT		COMPLEXITY	SIZE/QUALITY	DISPOSITION
HI	HIGH			
	MED			
	LOW			
MED	HIGH			
	MED			
	LOW			
LO	HIGH			
	MED			
	LOW			

CONTEXTUAL HINTS FROM KOREAN ANALYSIS

–Use of tactics against some threats
–Possible resource constraints may force more capability-based tactics
–IADS vulnerable to jamming
–Ground communication not vulnerable
_Ground-to-air communication vulnerable
–Sites netted via hardened ground line
–Can overload certain geographic sectors.

CHOICES

–IADS/SEAD campaign indicated to protect aircraft over battlefield
– –Releases close air support aircraft
– –Destroys North Korean "air deniability"
– –Enhances friendly air superiority
–Mixed tactics indicated
– –Not enough resources for sortie-by-sortie defense of mission packages
– –Destroy what cannot be jammed
– –Deceive what cannot be jammed or destroyed
– –Jam what does not need to be destroyed
–Phased use of resources baesd on what is available indicated

Figure 23. JSEAD planning matrix.

high-disposition array. It may even be possible for the intensity to change from one element to the other. The assessment of intensity is often dictated by "higher authority" (NCA) or by circumstances that call for a specific level of effort. For instance, a "low-intensity conflict" with "high-size/quality" may require "high-intensity conflict, high-complexity" JSEAD measures, depending on the guidance the theater commander receives and passes on to the battle staff.

We will use North Korea as an example of how the matrix might be used in a war-fighting situation. The North Korean IADS is well integrated and heavily redundant. It is based on a 1950s to 1960s-style Soviet IADS and not nearly as sophisticated as the Iraqi IADS. We could assess the complexity as medium when compared to other IADS systems worldwide. If the North Koreans were to upgrade to more advanced SAMs or surveillance radars, the IADS' complexity could easily be assessed as high. Additionally, the North Koreans do not place heavy emphasis on radar-directed organic air defense for their offensive ground forces, apparently preferring to extend their strategic IADS infrastructure to cover the geography in which their offensive operations will occur.

The North Koreans possess large numbers of air defense weapons and maintain them in good order. The density and overlapping coverage of their systems is excellent (i.e., destruction of a single system will not harm the IADS significantly). Thus, the North Korean IADS can be assessed as "high" in the size/quality index.

The North Korean IADS net is disposed countrywide, especially around their major industrial centers and along their borders. They have placed large numbers of their most capable systems along obvious avenues of approach to targets they consider valuable. The North Korean IADS disposition reveals a sensitivity to protecting urban industrial complexes and detecting attack from all directions early (at the expense of spreading these assets thinly). Aside from the obvious high-disposition index indicated, this assessment reveals a potentially critical weakness in their IADS structure based on Soviet doctrine: degradation of their early-warning system through jamming or selected destruction could easily cripple their air defense net along selected avenues of approach. We could assess the North Korean IADS, then, as one of medium complexity, high size/quality, and high disposition.

The intensity of a potential conflict is generally a "given," at least for planning purposes and during the initial phases of hostilities. Nonetheless, the characteristics of the intensity are helpful to JSEAD planners. The decision to use overwhelming offensive force, for instance, allows the JSEAD planner greater latitude in planning "phases" of the campaign. The virtually simultaneous accomplishment of the four phases of the Desert Storm air campaign (of which JSEAD was a phase) was achieved by the possession and use of overwhelming force.[12] In retrospect, SEAD planners discovered that there were great synergies created by the simultaneous attack on the enemy's C^3, strategic infrastructure, and IADS. JSEAD emerged as a critical element in enabling the air campaign's success.

The synergy created by JSEAD and the strategic attack on the Iraqi military infrastructure enabled a numerically equivalent force (ground units and total military personnel) to project crushing military power far out of proportion to its relative size. JSEAD possesses the curious property of being able to turn a "sufficient" force into an "overwhelming" force. If the enemy is defenseless from aerial attack, *any* aircraft capable of bringing weapons to bear against the enemy's unprotected centers of gravity represents an overwhelming force.

The best way to use JSEAD is to focus on objectives, not resources. Doing so does not make the question of resource availability, quality, or quantity immaterial; it merely places these factors in a perspective which subordinates them to overall objectives. There will be situations where the joint force commander will have to fight with limited resources and other situations where the commander will be able to acquire whatever is needed. *It is impossible to overemphasize, in either case, the requirement that all criteria of effectiveness be related to attainment of overall objectives*, and JSEAD, as a primary tool in attaining and keeping air superiority over the battle area, is no different in this respect.

Notes

1. A simple number and line plot from *Jane's Weapon Systems* (London: Jane's Publishing Company, 1972–73 forward) shows an almost geometric growth rate among command-and-control systems, early warning radars, and acquisition systems. Not only are the individual systems proliferating but the combinations and permutations of the systems are growing as well. To develop specific weapons systems to counter such a proliferation of numbers and types of future systems seems an imprudent and financially ruinous course to follow.

2. The author was a member of the Operation Just Cause planning staff and participated in the planning of the electronic combat portion of the operation.

3. Gen Michael J. Dugan, USAF, Retired, "Air Power: Concentration, Responsiveness, and the Operational Art," *Military Review*, July 1989, 21.

4. Interview with Col John A. Warden III et al., Headquarters, USAF, Washington, D.C., 31 July 1991. As mentioned earlier, Colonel Warden expressed surprise that the Desert Storm EC planning cell was able to "ramp up" so quickly when compared to other portions of the staff.

5. Derived from Philip A. Crowl's "The Strategist's Short Catechisms: Six Questions without Answers" contained in *Military Studies Course Book 1* (Maxwell AFB, Ala.: Air War College, 1991), 180–85. The questions are:

    1. What is it about?
    2. Is the national military strategy tailored to meet the national political objectives?
    3. What are the limits of military power?
    4. What are the alternatives?
    5. How strong is the home front?
    6. What have I overlooked?

6. Dugan, 21.

7. Joint Publication 1: *Joint Warfare of the US Armed Forces* (Washington, D.C.: Joint Chiefs of Staff, 11 November 1991), 21.

8. "Tactical Bombing of Iraqi Forces Outstripped Value of Strategic Hits, Analyst Contends," *Aviation Week and Space Technology*, 27 January 1992, 62–63.

9. *JSEAD: Multi-Service Procedures for the Joint Suppression of Enemy Air Defenses* (Langley AFB, Va.: Headquarters Tactical Air Command, ALFAs 1990), 2.

10. Ibid.

11. The "SEAD" package formations used in Operations El Dorado Canyon, Just Cause, and Desert Storm varied from the simple use of jamming aircraft to suppress radars as a separate formation with area fighter protection cover, to an elaborate series of jammers, HARM-shooters, Wild Weasels, special forces, F-117 bomb droppers, and fighter-bombers carrying cluster bombs to suppress poorly defended radar sites—all as part of an overall SEAD campaign to neutralize the enemy's total capability to use air defense.

12. Interview with Colonel Warden.

Chapter 5

JSEAD: Strategy, Tactics, and the Changed Threat

This chapter explores the opportunities that the changing world provides for developing joint SEAD strategies and tactics. The changed nature of the world's military environment has enormous implications for the role an enemy IADS will play in any future conflict. The changing nature of the American military provides both opportunities and limitations that developers of JSEAD must keep in mind when formulating plans to deal with the increasingly sophisticated IADS threat. In dealing with JSEAD strategy and tactics issues, this chapter discusses how a joint approach to SEAD tactics can reshape the way we think about and fight emerging enemy air defense structures.

The enemy IADS of the future will undergo fundamental changes. Universal access to advanced technologies will accelerate the pace in leveling the SEAD battlefield. While fiscal constraints slow the pace of US technological growth in SEAD equipment, the rest of the world—to the delight of electronics and arms entrepreneurs worldwide—has the opportunity to catch up. Ironically, the forces that controlled the proliferation of advanced electronics technologies during the cold war are evaporating with the demise of the Soviet Union. The US "edge" in SEAD combat, therefore, can no longer rest upon superior technology as the key element in achieving overwhelming advantage. The US edge—if it is to be maintained—will be held by integrating existing SEAD resources and combining the joint skills of all the services' SEAD expertise.

The evolution of SEAD paints a pattern of convergence—a convergence that will eventually make JSEAD a functional reality. JSEAD for US forces does not exist yet because each of the services' doctrinal, strategic, and tactical applications of SEAD has prevented it. This observation is not pejorative; functional joint warfare has not evolved far enough to allow an application of doctrine, strategy, and tactics that is universally shared by the respective services. Nonetheless, convergence among the various services' views on SEAD application is a prerequisite if joint SEAD is to become a reality. Fortunately, evolutionary forces are propelling SEAD strategies and tactics in that direction already.

Operation Desert Storm, while it applied each of the services' assets in a joint SEAD campaign, did not suffice as a true JSEAD venture. Individual service paradigms of warfare and tactics were applied to achieve joint objectives. The Desert Storm SEAD campaign was a very cleverly crafted amalgamation of single-service SEAD assets and tactics to achieve overall goals. The Air Force-dominated electronic combat cell that directed SEAD operations created a quasi-joint SEAD strategy where all services worked through the existing Air Force electronic combat structure. This was especially true for Air Force EF-111 and Navy/Marine EA-6B radar jamming operations. Differences between Navy/Marine F/A-18 and Air Force F-4G capabilities and basic tactics, however, prevented similar uses of these HARM-shooting assets at the tactical level.

Future JSEAD will differ significantly from Desert Storm applications because individual joint war-fighting doctrines, strategic paradigms, and tactics are converging. For a variety of reasons, the basic capabilities of service-held SEAD assets and tactics are becoming more similar. The F-4G will probably go out of service soon. The military budget cannot afford a replacement for the kind of capability the F-4G represents. Future lethal SEAD capability will reside in HARM-shooting aircraft that will have some of the capabilities of the F-4G, but Wild Weasel tactics in the joint arena will change. The institution of a single information-processing net linking all the services on the battlefield in real time will also allow a more fluid use of each of the services' assets as a function of overall *joint* objectives rather than service objectives. The movement of evolutionary forces towards convergence not only aids the transition from the quasi-joint SEAD strategies and tactics of Desert Storm to true joint SEAD, it establishes the ground work for it.

Four major evolutionary forces contribute to the movement towards convergence:

1. The uncertainty of the future. Planners do not know what to plan for. Who is the threat, where is the threat, what is the threat, and when is the threat? The range of uncertainty circumscribed by these questions is pushing all services toward general capability-based concepts that are "good enough" to meet the threat as a whole, but not good enough to meet the threat individually.

2. The nature of the future threat's air defenses. The availability of sophisticated and relatively inexpensive air defense systems by nations inimical to US interests is growing. The mercantile instincts of the world's arms industries promise to accelerate this growth.

3. The rapid dwindling of SEAD combat resources. The traditional threat to Western survival has diminished, but, curiously, none of the conflicts in which US SEAD assets have been employed have been against the central threats for which they were designed. Ironically, regional conflicts where SEAD assets have been employed promise to continue, and the enemy air defense systems promise to grow as a function of what is available on the open market. We have reached the limits of doing "more with less"; future JSEAD will have to acknowledge these limits as a function of pooling resources and being smarter in the "more-with-less" game.

4. The relationship between forward-deployed forces and US-based power-projection forces. The requirement to respond quickly and decisively anywhere in the world against potential enemies of varied air defense capabilities places an absolute limit on what US forces will be able to do to the enemy, especially if American forces must employ military force directly from the continental US. By forging common links through mutually understood and thoroughly practiced JSEAD, US forces will be able to overcome weaknesses that individual service weapons and tactics might not be able to. Global reach, global power, the JSEAD way, may presage a novel way of packaging the available forces.

Uncertainty, diminishing SEAD capability against a relatively more capable SEAD threat, and a changed basing posture for US forces compel the movement toward sharing SEAD resources. This pressure has raised penetrating questions about SEAD force structure, organization, composition, command and control, authority, and unity of effort. For SEAD practitioners, these questions are well on their way to being answered. However, the general inertia associated with institutional change is slowing the process. Vested institutional interests in each of the services has dampened the evolutionary movement of SEAD. The service cultures lag behind the incipient movement toward "true" JSEAD. Unfortunately, the answers to the JSEAD questions cannot become reality until the institutional vested interests overcome this inertia.

Not all evolutionary trends will enhance the transition to JSEAD strategy and tactics application. Part of the problem concerns the differing foci of strategy and tactics. Unlike SEAD *strategy*, which deals largely with objectives and concepts, SEAD *tactics* deal with physical reality and pragmatic application. The strategic requirement to maintain a general capability-based force runs headlong into the equally strong tactical requirement to meet specific objectives. To SEAD tacticians, the IADS threat exists as a physical entity in a specific place with specific properties and capabilities, and the IADS threat is netted together in very specific ways. The enemy personnel who operate the air defense systems are trained to employ their systems with a specific doctrine and are commanded by leaders who operate through a specific organizational structure. These "specific factors" make up the world of the SEAD tactician. The problem is that American SEAD tacticians no longer know ahead of time *which* specific systems to plan for.

As air power continues to gain importance as a principal element of power projection for each of the services, *JSEAD tactics* will become more and more important at the "nuts-and-bolts" tactical level. The problem for SEAD strategists in adapting capability-based, resource-scarce forces to meet the threat is compounded because US forces no longer know who—or exactly what—the threat is. SEAD strategists must plan for more general contingencies. Tacticians, on the other hand, must configure existing forces and weapons in employment schemes that counter a specific—and likely undefined—threat. The paradox is daunting. The traditional paradigm of pairing a specific friendly weapons system against a specific enemy target set or weapon capability is no longer tenable; yet when war comes, the US "general-capability SEAD force" must become specific enough to defeat a unique—and specific—enemy IADS threat.

The requirement to counter specific threats with specific weapons and the fact that US forces will have fewer resources with which to counter these threats poses a classic dilemma. We pit the resource-scarce "general" force against the tactics-driven "specific" threat. On the one hand, SEAD forces must be able to deal with the specific threat effectively enough to accomplish the combat objective and maintain sufficient survivability. On the other hand, because resources are scarce, the SEAD resources themselves must be flexible and versatile enough to perform as many functions as they realistically can.

Clearly, if JSEAD tactics are to be applied in any meaningful tactical way, we must develop a viable solution to the dilemma. Since the IADS threat is the primary driver in JSEAD strategy and tactics formulation, we must first understand its nature.

The Essence of the Threat

Paradigms change because the world changes. When it is obvious old ways of looking at reality no longer fit, new ways must be invented. The same reality that compels American single-service SEAD strategies to evolve into JSEAD strategies also compels our potential enemies to evolve. IADS structures with which American forces must deal have undergone radical change in response to the changed world. The transition in concepts from a linear to a nonlinear battlefield and from a vertical, sequential decision-making process to one using a cybernetic model also applies to the potential enemy.

The principal factor altering the nature of the threat posed by the modern IADS is the treatment of data and information. As a direct descendant of the technological revolution, the introduction of information and data-processing networks has transformed the modern IADS.

The transition from technology-based to information-based IADS has occurred rapidly. The transition is traceable through three distinct phases. Each of these changes flows from the way technological innovations have been applied to information processing technology. Through these changes, the modern IADS has undergone an unwitting transformation where information processing seems to have supplanted the technology that spawned it.

The first phase of the transition occurred when radar was mated with the analog computer. The second phase occurred when more advanced radars were netted to a central system of command and control made possible by more advanced analog computers. The third phase was (and is) the introduction of digital computer technology and its application to radar technology as well as to information processing. The digital computer enabled major qualitative advances in computer design and radar technology and gave real impetus to the creation of the information-based modern IADS.

In the Vietnam era, systemwide information gathering and distribution played a secondary role to the physical ability of individual systems to destroy invading aircraft. North Vietnamese SA-2 sites, for instance, were initially responsible for acquiring and tracking their own targets. Available computer and radar technology limited the ability to coordinate a systemwide information gathering and dissemination net.

With the advent of computer technologies that could link detection systems to provide timely warning and accurate tracking data, information gathering and dissemination began to play a more significant role in the application of air defense weapons. Gradually, command and control assumed a greater role in the use of air defense assets by channeling information to desired locations. This was especially apparent during the Israeli use of SEAD in the Bekaa Valley. Computer technology formed the basis of the system, and that technology was still limited in the amounts of data it could process and transmit in a timely manner. The demand for data was still greater than the technology could produce.

The IADS/C^3 systems used prior to the introduction of digital computer technologies were still linear because information was restricted by the technology available to a generally sequential flow. Information traveled a specific path. From the early-warning radar, detection information went to a filter center. From the filter center it was relayed to a command center. The command center sent a decision to the filter center and fire control site. The fire control center acquired the target and shot it down. The total air-situation picture was not available to the early-warning radar site, the filter center, or the fire control director. It was even difficult to acquire a complete air-situation picture at the command center because the data presented tended to be a mixture of information gained over different segments of time. The communication path used by this IADS/C^3 generation was iterative and sequential; consequently it tended to be time-consuming and cumbersome.

With the advent of digital technologies, most prominently displayed in the Iraqi IADS, the information flow became nonlinear. The inherent makeup of the digital computer-based system made huge amounts of information immediately available to any part of the IADS on demand. Any part of the system had the technological capability to access the information any other part of the system possessed. Feedback loops were *part of the technology,* not a process controlled by decision makers. Information accessibility dominated the entire IADS/C^3 structure as an unwitting by-product of the technology itself. The application of digital computer technology catapulted information processing into the preeminent role it now plays. The simultaneous transmission of pertinent data throughout the entire IADS array makes it virtually "nodeless" in the classical targeting sense.

Figure 24 shows the rapid transition from shooter-based to information-based IADS. The relationship between technology and the evolution of the IADS, while undeniable, tells only part of the story. The evolving air defense system's technological capabilities met—and then rapidly exceeded—the paradigms that generated their creation. The multiple uses of digital computer technology and the myriad permutations resulting from netting the thousands of communications, radar, and air defense weapons available together by software makes the concept of a single SEAD weapon for a single IADS threat obsolete.

The modern IADS is multilayered and linked through an intricate, redundant command-and-control structure. It is heavily information-dependent. Because the IADS is heavily dependent on information, *information denial* becomes a key ingredient in modern IADS suppression.

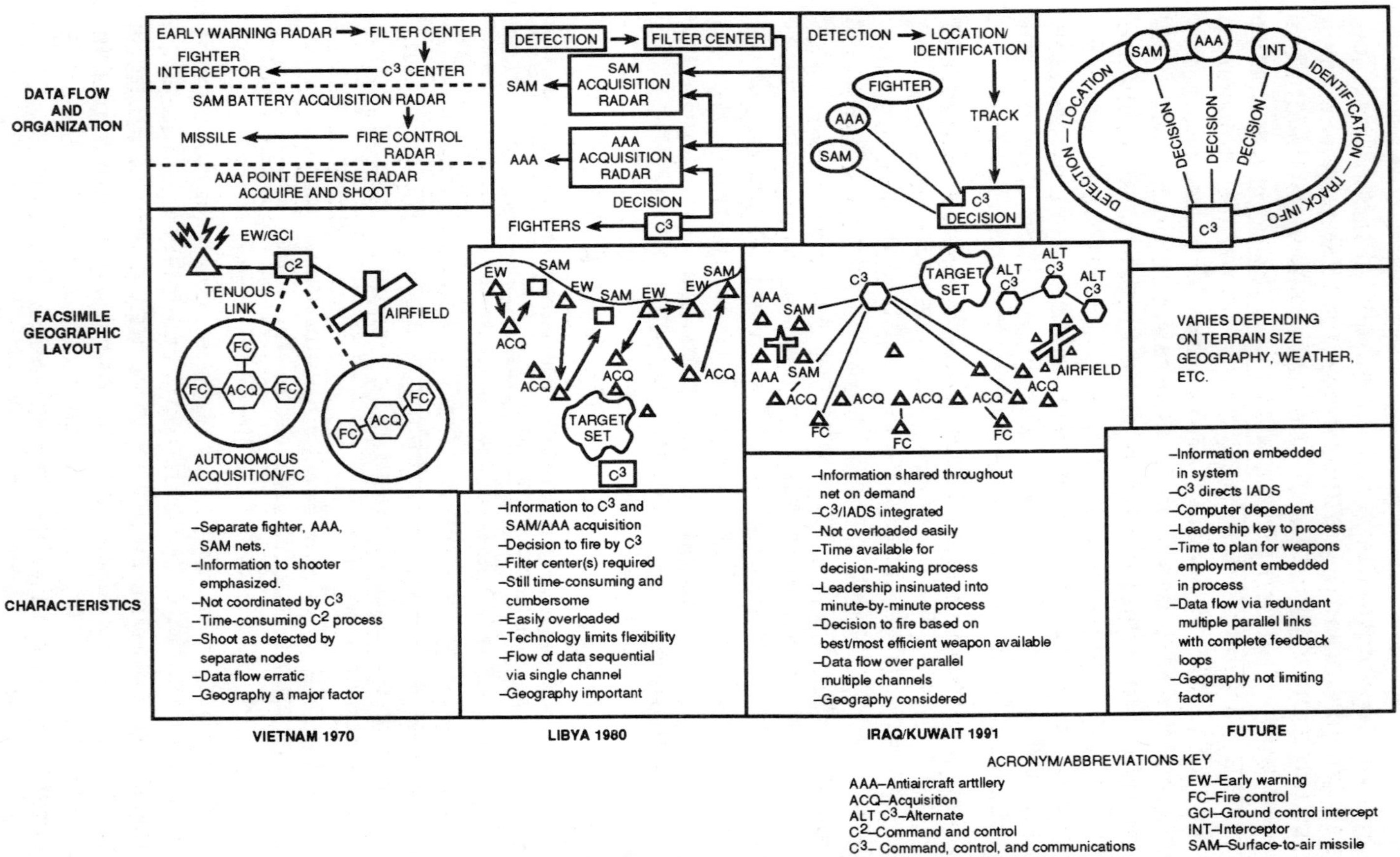

Figure 24. Evolution of information-based IADS.

Figure 24 shows that an exchange of information takes place at every point in the modern IADS. This has not always been the case. *Classic* SEAD strategy focused on the United States' ability to destroy key points of the IADS, preferably the shooters. Current SEAD strategy, while noting the importance of applying a wide array of disruptive and destructive tactics to the enemy IADS, has not departed significantly from this paradigm. The reason for this is pragmatic: there is nothing so comforting to the warrior as the physical destruction of enemy forces—even when disruption may be all that the situation requires.

JSEAD must evolve beyond the classic "destroy and jam" strategies of resource-rich times when US SEAD technology, personnel training, and firepower outstripped virtually any opponent. In adapting to the IADS's transition from shooter-based to information-based orientations, JSEAD must focus increasingly on *information denial* as its primary task. The unwitting transition from a shooter-based to an information-based IADS provides modern SEAD strategy with the key vulnerability of the modern IADS structure: paralyze the enemy by denying the ability to acquire, process, or transmit data.

In hindsight, Operation Desert Storm marks the first time a SEAD campaign based on information denial was undertaken. Ironically, SEAD strategists and tacticians, when analyzing key elements of the war, discovered *after the fact* that information denial was the principal focus of the SEAD campaign. It is clear from the sequence of observed events that the *initial* phases of the SEAD campaign were directed at the Iraqi IADS/C^3 net's ability to acquire, process, and transmit pertinent data (information-based). The follow-on objective was suppression of AAA, SAM, and Iraqi interceptors (shooter-based) through a mixture of intimidation, confusion, and occasional physical destruction, but—from an overall strategic view—the shooter-based view was subsumed as a function of information denial. One of the primary differences in SEAD application between Desert Storm and previous military operations was the discovery that *disrupting or denying data flow—without ever physically challenging the shooting apparatus of the air defense network—cripples the modern IADS.*

The modern IADS—with redundant communication nodes and instantaneous access to data—presents an almost impenetrable barrier to Vietnam-era SEAD strategists. Viewed, however, as an evolved information-gathering network attached to air defense weapons, the modern IADS reveals its key vulnerability. In order to acquire the speed and efficiency the total net demands, the modern IADS has unwittingly become *dependent* on electronic information gathering and dissemination; it is as effectively neutralized by information denial as it is by destruction.

The technological revolution in IADS equipment design—from radar-based to communication-based networks, from simple radar-guided AAA to "fire and forget" modern SAM with on-board, terminal-homing radars—has pushed the evolution of the modern IADS faster and farther than most other aspects of warfare. This asymmetrical evolution of the IADS threat has also propelled SEAD strategies and tactics to meet it. Because of this asymmetrical evolution, the tactics supporting modern JSEAD operations may not seem to make sense in relation to tactics that support other mission areas. Some aspects in the evolution of modern JSEAD tactics may even appear bizarre or contradictory when viewed from more traditional air power tactics perspectives.

Speed is the essence of the modern IADS threat. While modern aircraft penetrate enemy air defenses and fly over the battlefield at speeds measuring in the thousands of miles per hour, the modern IADS reacts at the speed of light. The future IADS will provide the enemy a virtually instantaneous picture of the aerial situation. This picture combines the traditional IADS responsibilities of detecting, locating, identifying, tracking, allocating weapon systems, and issuing firing instructions, but it accomplishes these tasks as part of a very rapid, fluid process. The difference is more than speed and amount of data flow, more than streamlined command and control; it represents a qualitative leap in capability for the modern IADS.

Modern IADS structures available on the world market—of Warsaw Pact origin, Western, and (increasingly) indigenous manufacture—have track capacities (the electronic computer "tag" given an aircraft that has been detected, located, and is being tracked) in the tens of thousands. The overall picture advanced information-processing systems present can be manipulated to focus on a specific area—or a larger area—on demand, and this picture can be transmitted instantly to any location within the command-and-control net.

The redundancies offered by conventional telephone cable, coaxial cable, or fiber optics, in addition to different types of highly directional microwave and radio links, make targeting the modern C^3 structure an ambiguous task. Because the system is inherently nodeless, it is difficult to know what constitutes a "critical C^3 node" or how to determine where the critical node is. Even defining what constitutes destruction is a difficult task for the SEAD/C^3CM tactician. "Bombs on target" no longer carries the certainty of desired system destruction it once did.

The speed with which a modern enemy IADS adapts to attack poses some special problems for the intelligence community—especially as it relates to collection, assessment, and dissemination. Collection equipment must be able to pinpoint emitters and data transmission arrays in real-time. Assessment of enemy intentions and capabilities must be quick, accurate, and disseminated rapidly to SEAD strategists, tacticians, and users.

The SEAD community's intelligence requirements have changed in direct response to the qualitative changes in the modern IADS. Tacticians must have a clear picture of enemy information-processing capabilities as well as weapons characteristics. Because the modern IADS can adapt so quickly to changing circumstances, it can present a chameleon-like target. This ability to adapt quickly places stress on an intelligence system that was not designed to collect, analyze, and disseminate pertinent data to US SEAD tacticians at the speed with which the modern IADS changes.

Timely data—real-time data—is crucial to the development of a clear picture of the enemy net. Relevant day-to-day tactics development and targeting priorities require rapid feedback and analysis to be effective in the information denial campaign against the modern IADS, and it is uncertain whether traditional intelligence collection means and processes are up to the task. The rapid evolution of the modern IADS seems to have caught the intelligence community—usually one of the stalwart supporters and enablers of the SEAD community—flat-footed. Until this shortcoming is corrected, the intelligence community's limited capabilities with respect to the modern IADS will have a constraining effect on JSEAD.

The IADS' transition to an information-based system has also created changes in its requirements. Examining these information-based requirements—and the way the modern IADS treats them—will reveal some lucrative areas for exploitation.

Requirements of the Information-based IADS

To bring air defense weapons to bear, an enemy IADS must be able to detect the presence of attackers, locate and identify them, establish a record of the attacking aircrafts' tracks, and finally, allocate and employ air defense weapon systems against the attackers. The process itself remains essentially unchanged since World War II, but the volumes of information and the speed with which the process operates have made qualitative differences in the capabilities of the modern IADS. Examining how the detection-to-weapons-employment air defense procedure has evolved—through the lens of information-processing technology—reveals the extent to which the modern IADS has been transformed.

Detecting

Traditionally, visual spotters and spies were used for detecting enemy aircraft. They are still used, but as a complement to electronic means of detection. Electronic means are both active and passive. Active radars emit a characteristic waveform to detect aircraft at great range (150 to 300 nautical miles, depending on atmospheric conditions and line of sight). However, what constitutes a "characteristic long-range radar waveform" is changing rapidly. Some newer radars do not even have to worry about line-of-sight problems. Bistatic radars, for instance, use one site to bounce radar waves off the ionosphere and the other to receive the data. By comparing the data at transmission—the known characteristics of the ionosphere and the received signal—a prospective enemy might be able to detect aircraft well beyond the horizon. Other modern radars radiate signals spread over a wide frequency spectrum and use stretched-out pulses (which are very difficult to jam) to clarify the radar picture at extreme ranges.

The most troublesome electronic long-range detection "threat" of the modern IADS, however, is the passive receiver. It is troublesome because friendly aircraft have no way of knowing if they are being tracked passively. Worse, because passive detection sites do not radiate, it is extremely difficult to identify or locate them. This makes the traditional point "jamming/destroying" tactic unusable. In fact, the attempt to jam or destroy such a site may be exactly what a passive array needs to locate and identify intruders.

Not only do airborne radars and radios emit detectable electromagnetic radiation, so do computers in modern aircraft—albeit at very low power levels. Passive detection arrays may have evolved to the point where they are sensitive, accurate, and persistent enough to locate and identify aircraft at long range. They may come to supplant the radar as the primary means of detecting ingressing friendly aircraft. They may even become accurate enough to aid in the locating, identifying, and tracking portions of the IADS process. At the very least, adding passive detection arrays to an enemy IADS net would complicate the SEAD planner's job.

The modern IADS has many countermeasures that do not rely on sophisticated technology. Possessing large numbers of radars, using varied frequencies and different radar waveforms of various complexity, turning radars on and off to confuse the intelligence collection effort or complicate targeting—all of these enemy countermeasures have been reasonably effective in the past. Even in the transition to a SEAD strategy based on information denial, these countermeasures will still pose problems to the SEAD planner.

Locating and Identifying

There is a considerable difference between knowing that something is out there and knowing where and who it is.

Detection need not be precise; it need only determine if something is out there and about where it is. Locating and identifying a target is a more complicated and precise venture than detecting one. The radar waveforms used by location and identification systems also are changing rapidly because of technological innovation. Current waveforms used in locating and identifying vary from the type used in early-warning detection radars to those employed by short-range acquisition radars.

The important distinction between the detection and location/identification phase is the treatment of the information. Detection energizes the IADS structure to prepare for the intruders. Location and identification data is more precise and there is more of it per time bit. The IADS structure transmitting this data requires a system that can handle the increased volume at higher rates. A key characteristic distinguishing the modern IADS from its earlier forms is its ability not only to locate and identify intruders with very capable radars and passive arrays, but to transmit the high volumes of data required at the very fast rates required by decision makers. The capability to locate and identify targets in the information-based IADS is important only to the extent that the system is capable of transmitting the information to the appropriate places in the IADS decision-making apparatus.

The type of equipment responsible for the location/identification of detected aircraft is in a state of radical redesign. Digital computer technology makes it possible for some radars to customize their own waveforms to suit the task. The advanced radar/digital computer combination—like the tracking system supporting the American SAM-D Patriot surface-to-air missile—is capable of altering its radar characteristics thousands of times a second to accomplish the tasks set before it by the Patriot's computer software programming. More importantly, the radar can share its information with other radars via data link or receive enhanced data from long-range exterior sources through the same link. Patriot-like systems—systems that meld advanced computer technology to advanced radar and data transmission design—are apt examples of the kinds of systems employed by the location/identification phase of the information-based IADS.

The problem this poses for traditional SEAD strategy is that it is difficult to distinguish where the location/identification function ends and the communications function begins. The problem arises because in traditional SEAD strategy, the weapons and tactics choices employed against location/identification systems and the C^3 structure tend to be separated as if they were dissimilar entities. In the modern IADS, C^3 nodes, radar nodes, weapons nodes, and location/identification nodes are becoming indistinguishable.

The enemy's requirements for location/identification equipment also create a curious technology problem for the SEAD planner. The availability of differing technologies has been complicated, ironically, by the removal of a common enemy technology to which the US SEAD community had grown accustomed. The familiar lexicon of electronic combat and SEAD has been transformed by the demise of the Soviet Union. The classic early-warning, acquisition, and fire-control radar waveforms of the past, while still the location/identification mainstay of many potential enemy IADS, are rapidly being replaced with newer generation radars—many of mixed Western and Soviet design.

The change is qualitative; the location/identification systems of the modern IADS have allowed potential enemies to leapfrog ahead in technologies. The radar and communications systems that future US forces may face will likely be hybrid melds using state-of-the-art complex radars similar to the SAM-D Patriot. These radars—unlike those of cold war-era potential enemies—will probably be manufactured by companies such as Raytheon, Westinghouse, CSF Thompson, Marconi, Plessey, or Siemens and be supported by the most modern communications technologies marketed.

As complex and advanced as location/identification radar technology has become, it is pushing the limits of what it can do physically. The next quantum leap in locating and identifying attacking aircraft for the enemy IADS is occurring right now in networking. The broad applications of digital computer technology to information processing open vast possibilities at reasonable cost to the potential buyer of a modern IADS. Relatively small, inexpensive computers—complete with high-resolution displays and high data-rate transmission mechanisms—are being netted together through master computers that can interpret data from radars, passive detection arrays, visual observation posts, and spies; these systems can even suggest weapons allocation options to the battlefield commander. The new information systems are capable of netting old radars with new ones, combining old systems into a larger array, or revitalizing previously existing systems. The revolution in information systems qualitatively changes the capability of the IADS net, and that revolution is just beginning. This gives a disquieting tone to the fact that it was a 1950s-vintage Soviet SAM netted to an Iraqi-designed, hybrid communications net using both Western and Soviet equipment that brought down a US Navy fighter during Desert Storm.

What does all of this mean to the SEAD planner? First, there is an absolute physical limit (meaning "set by the laws of physics)" to what current friendly SEAD weaponry can do. There is a proliferation of advanced radars on the open market designed to locate and identify aircraft without the capability to detect and target such radars. Second, the evolution from traditional pulsed radars as key nodes of a crude information net to modern radars reliant upon an elaborate network driven by digital computers has produced a key IADS vulnerability.

The new location and identification segments of the modern IADS network, though more resistant to traditional forms of SEAD and electronic combat, are perhaps more vulnerable than the older system in a number of ways. First, computer processors are vulnerable to overloading. For example, one needs only find a way to force the processor to process the data it considers information but which is, in fact, something else. Microchip processors are "bit counters." So long as the software program that directs the processor to count bits remains in force, the processor will count until it burns out. The challenge to the SEAD tactician is to determine how to trick the software into believing the data bits counted by the microchip processor constitute "real" data.

Second, almost all computer-driven systems present synthetic displays. If the computer program does not recognize the patterns the radar is sending, it will not display them, or it will display them incorrectly. This provides the SEAD planner an opportunity to employ the most effective type of electronic jamming available: jamming that is undetectable to the enemy.

Third, computer-driven systems have not evolved to the point where they are hardened to environmental hazards such as rain, dust, and extreme temperature or humidity changes. They tend to break down more often than older systems in inclement conditions and, from a tactical point of view, provide conditions that are more opportune for attack.

Tracking

Tracking is the single element that computers and modern networking have aided the most. Electronic tracking in a modern IADS has eliminated the requirement for scores of people to move figures around a giant situation table at command centers and for specially trained personnel to update ungainly greaseboards to keep the battlefield commander informed. Modern tracking techniques have also eliminated middle-level filter centers because there is no longer the requirement to screen the data coming to the command center. All data can be transmitted immediately to the command center as soon as the locating and identification system has tagged it. Modern tracking techniques not only provide enemy commanders a real-time, dynamic picture of the air war, they identify which airplanes are going where and what air defense weapons systems are available to counter the attacking aircraft.

Once the enemy IADS has detected, located, and identified the target, the IADS system must keep track of attacking aircraft to allocate available air defense weapons to them. Modern computer technology enables the detection net to determine if a formerly detected aircraft is the same one another system sees or a different aircraft. The tracking system assigns a track number and refreshes the data periodically, depending on the priority the tracking system has assigned the target. Higher priority targets have a more frequent data-refresh rate than those with lower priority. The enemy IADS' tracking system assesses what air defense weapons are available to counter attacking aircraft. It may then select the appropriate type and number of weapons to ensure a reasonable probability of kill against the target. The electronic display of the IADS tracking systems presents this data to the battlefield commander for a weapons commitment decision, or, if the air defense net is set to automatic operation, the tracking system will allocate the appropriate weapons to kill the tracked target according to its programmed priorities.

Weapons Allocation and Employment

The US aviator is more interested in air defense weapons allocation and employment in the IADS process than in any other. This is understandable because the aviator's stake takes on physical meaning at this stage of the process. This is where the US aviator meets the enemy IADS face to face: it is no longer the unseen maze of electronic equipment and electromagnetic waveforms, it is a personal threat. If the modern IADS completes a successful transfer of data to selected air defense weapons, then aviator survivability will be low. Ironically, it is the unseen portion of the modern IADS process prior to air defense weapon allocation and employment that is responsible for the increased lethality of the modern IADS array. The persistent refinement of data on the US attacker's air speed, altitude, location, type of aircraft, and probable target allow the enemy the luxury of time to determine how to eliminate the attacker. It allows the enemy IADS to take the initiative by choosing the time and place to employ weapons.

The modern IADS capacity to allocate and employ its air defense weapons is related directly to its information-processing capability. This makes the unmolested modern IADS more lethal to the aviator in the weapons employment phase of the IADS operation because the enemy system has more time and information available to prepare a kill zone for the attacker. What makes this situation especially anxious for the aviator is that there is no way to know if the enemy air defense weapons engaging the aircraft are the end product of a fully coordinated process or a blind, last-minute desperation shot by a frustrated enemy weapons operator.

The weapons the enemy can bring to bear will be well suited to the information-sharing environment. Table 10 provides a brief overview of the type of weapons that the future US attacker will face. The systems available will cover every niche of the battlefield. The ranges indicated show the general environment in which the air defense weapons were designed to operate—not the system limitations.

Table 10

Traits of Modern IADS Weapons

Long-range systems (50–100 nautical miles)
- Multiple target track/multiple simultaneous launch
- Track via missile (on-board radar) to enhance low-altitude capability at range
- High speed, high maneuverability
- Resistant to jamming countermeasures
- Highly dependent on computers
- Very complex
- Identification friend or foe a problem
- Very expensive
- Few in number
- Not very mobile
- Usually a strategic system
- Not very vulnerable to antiradiation missiles
- Designed to present ambiguous picture to radar detection equipment

Medium-range systems (10–50 nautical miles)
- Large numbers available
- Large variety from old and unsophisticated to new and very sophisticated
- Most are targetable by antiradiation missiles at standoff ranges beyond lethal envelope of defense system
- Most are detectable by Western electronic warning equipment
- Most have excellent low-altitude capability (below 500′)
- Many are mobile, can be set up in new locations in minutes
- Many have multiple-mode capability to include radar, optical, and infrared tracking/guidance
- Dependent on radio communication in mobile mode
- Newer models extremely difficult to jam; most, however, vulnerable to jamming
- Reaction time critical to target acquisition for most medium-range systems
- Traditionally allotted to sectors; under new IADS command-and-control arrangements, can be controlled by central command; weapons-firing doctrine dependent on training, organization

Short-range systems (0–10 nautical miles)
- Designed primarily for point defense and organic air defense protection of field armies, navies
- Excellent low-altitude capability
- Almost all radar systems have optical or infrared backup, or both
- Some pure optical-, laser-, and infrared-only systems
- Extremely quick-reacting; but prior warning alert significantly increases probability of kill
- Usually late warning to aircrew that they are under attack

- Radar-directed systems vulnerable to antiradiation missiles, infrared systems vulnerable to being decoyed by infrared flares (so long as the intruder is aware of attack by an infrared system)
- No current countermeasure to optical- or laser-guided systems, other than maneuvering the aircraft once under attack
- Usually less than 10 seconds from launch to impact in the point-defense mode
- Extremely mobile, difficult to detect
- Autonomous and/or centrally controlled; more effective in autonomous mode, but will probably be used in centrally controlled manner because of nature of modern IADS C^3 structure

Fighter-Interceptors
- Modern fire-control systems netted to sophisticated GCI; all-weather/day or night capability
- Range dependent on warning time
- Accuracy dependent on ground-controlled intercept instructions and onboard fire-control systems
- Will likely consist of full range of modern fighters (to include F-16s, F-Is, Mirage 2000s, Fulcrums, Flankers, etc.) as well as full array of older aircraft
- Numbers and composition dependent on situation
- Training of aircrews likely to be deficient in comparison to US
- Doctrinal adherence, air discipline uncertain
- Tight GCI control will compensate for deficiencies in training, doctrine, and discipline

The air defense weapons of the future IADS make sense when viewed from the information-processing perspective. The technological revolution that created separate digital computer applications for radars, communications, and weapons systems has created another more ominous application of the digital computer. Instead of perfecting separate systems, the revolution has created a separate technology dedicated to melding the separate computer, radar, and data transmission technologies together. Information technology's aim is building devices that tie other systems together so as to exchange data efficiently. This makes the entire IADS process even more fluid and dynamic and presents SEAD its biggest challenge since Vietnam.

The contrasts between traditional SEAD strategies and tactics employed up to Desert Storm and those required to defeat the modern IADS are revealing in this regard. They reveal changes in the way we look at the battlefield, the timeliness of our responses to changes in the enemy IADS, and even the way we conceive of the IADS target.

First, traditional SEAD strategies and tactics are based on a division of labor which postulates a clear distinction of SEAD responsibilities by mission type. In traditional SEAD strategy, the offensive counter air (OCA), defensive counter air (DCA), close air support (CAS), battlefield air interdiction (BAI), and interdiction missions are supported by SEAD. The distinctions among these missions is blurring. Both the nature of modern US equipment and the nature of the modern IADS dictate multiple uses for the same pieces of equipment. Consequently, it is no longer clear when an F-15E destroys a SAM site en route to bombing an enemy airfield whether it has performed an OCA and a SEAD mission, a SEAD-supported OCA mission, or a SEAD mission with two targets. However, the destruction of the SAM site as part of an enemy geographically oriented information-processing system makes more sense because it denominates the destruction of the SAM site in terms of the enemy system as it exists, not as the SEAD tactician would like to see it.

Virtually every mission described in AFM 1-1, *Basic Aerospace Doctrine of the United States Air Force*, is growing to encompass elements of other missions. SEAD is no different; as the IADS threat becomes more sophisticated and more integrated into the enemy's strategic structures, SEAD, as described in AFM 1-1, has also grown to encompass the appropriate areas. Therefore, it is not surprising that many Desert Storm SEAD missions were actually executed as OCA, DCA, CAS, BAI, interdiction, or strategic missions because facets of the IADS had grown to influence the total Iraqi military establishment.

Second, the speed with which the IADS can react, given the data and available weapons resources, makes most classic SEAD employment patterns obsolete. Classic SEAD patterns are geographic in orientation: clear *this* corridor, suppress *this* area, protect *this* formation to *this* target. The orientation of the modern IADS is systemic: the IADS will shift to meet the threat where it is with the totality of the resources that *the entire system* can bring to bear.

Third, old SEAD patterns were threat-system driven, not IADS-structure driven. The focus was to destroy the enemy's lethal capacity rather than to suppress the enemy's ability to bring the lethal capacity to bear. Old SEAD patterns of employment had the luxury of numbers and capabilities over very specific target sets. New SEAD patterns—both because of the nature of the threat and diminished resources—must concentrate on neutralizing the IADS as a function of degrading its overall ability to bring weapons to bear. Destruction of key critical nodes—if they can be located—will no doubt play a major role in this new pattern, but the primary focus can no longer be the destruction of IADS' shooters; it must be the degradation of the IADS information-processing network.

This brings us full circle. The objective of the new SEAD concept is to stymie information flow, thus paralyzing the enemy IADS. The site-specific, geographic orientation of "destroy and jam" classical SEAD strategies and tactics is neither cost effective nor strategically viable as a primary focus of SEAD strategy. The focus of JSEAD strategy should be objective-oriented, not resource-oriented. The irony of JSEAD is that, since Desert Storm SEAD resources were so successful, the likelihood of acquiring better weaponry for the future threat—perceived as diminished—has lessened. The US military may find itself in the position of fighting a superior air defense system with relatively inferior weapons. SEAD objectives may have to be pared down because US SEAD assets may not have the physical capability vis-à-vis the enemy's new IADS structure. This does not mean US SEAD will be powerless, but it does call for a realistic appraisal of what can and cannot be done to the modern IADS with weaponry we currently possess and are likely to possess.

Defeating the IADS: Information Denial

Applying what we know about the nature of the modern IADS threat—what has changed, how it has evolved—to strategy and tactics development reveals the modern IADS vulnerabilities and provides some pertinent insights. Interpreting the modern IADS process from detection through air defense weapon employment and applying the strategy of information denial will set the stage for some innovative tactics.

The evolution of the IADS net has made communication and information the key elements of its structure. While the weapons themselves are still very important, the process linking the net together seems to have supplanted the parts as the primary mover. This changes the SEAD calculus from one of attack on the physical air defense weapon to one of attack on the information gathering, processing, and transmitting aspect of the net. The information-denial approach also provides a different vantage point from which to view the enemy's air defense structure. Defeating the modern IADS is more than an attack on the technological base of the system; it is an attack on the doctrine the technology has forced the modern IADS users to employ.

Detection as Information

Curiously, SEAD—using Desert Storm as the latest milestone—seems to have employed the "information-denial" strategy more as a matter of intuition than forethought. Maj Gen John Corder, director of operations for Central Command Air Forces during Desert Storm and commander of the USAF Tactical Air Warfare Center (responsible for the Tactical Air Forces' electronic combat equipment and tactics), made the following return in response to this comment and question, "Iraqi radar-directed air defenses proved singularly ineffective. How much of that was attributable to EW [electronic warfare]?"

> Well, if you think electronic combat, not EW, I would say that it all was . . . because we went out and we did everything. We did SEAD . . . , we did C³CM and we had our own on-board self-protection EW. We set about in a very deliberate manner to take that thing apart as the first order of business, the price of admission. That's what you do. So we bombed all the operations centers, we jammed everything we could on the first day. We knew the jamming would be very effective early, but we knew that you couldn't rely on that for the whole war. So we went into a very aggressive campaign to beat up on all [the] EW GCI sites we could find (I'm talking about direct attack). . . . We sent A-10s out the first day and the A-10s just had a field day on a lot of EW GCI plants, which were essentially undefended, and just really blew them apart. So we took away much of the EW GCI that way. Of course Compass Call [EC-130H] was doing its thing in the command and control business to keep [those Iraqi assets] under control until we could bomb the communications facilities and the other stuff that they needed to communicate with. To me, it was a classic campaign, not really a lot different from those we practice in a microcosm out at Nellis during Green Flag.[1]

General Corder's comments are classic testimony to SEAD's evolution. Described is tactics-driven strategy at its best. Without ever identifying the strategic or doctrinal source, the statement cements tactics to objectives in a series of clear observations. First, the distinction between C^3CM, SEAD, and EW was incidental to getting the job done (the objective). Second, since the IADS was a connected entity, attacking any one of these aspects had an effect on the others. Third, the SEAD plan was a deliberate effort to take apart the IADS in a sequenced, iterative manner. Fourth, the choice between disruptive means (jamming, harassment, deception) and destructive means (antiradiation missiles, standoff ordnance, direct attack) was based on a logical criterion: what was most effective at the time and place compared to what was available to perform the task. Fifth, SEAD was effective early because its practitioners were prepared through exercises such as Green Flag. This lends credence to the proposition that today's SEAD (as part of electronic combat) provides the tactical models that might be used by other mission areas tomorrow.

The enemy's detection network is critical to kill-zone preparation time. Denial of the information will not destroy enemy ability to bring weapons to bear but will buy time. In some instances, however, this may effectively destroy the enemy's ability to bring weapons to bear because (1) the time it takes for the enemy to prepare for battle depends on timely detection, or (2) the speed with which friendly forces move is such that the enemy cannot complete preparations for battle prior to friendly aircraft attacks.

Information denial of the enemy's detection apparatus carries with it a whole litany of questions. How long is it necessary to deny the information? Is it more convenient, permanent, or tactically viable to jam, deceive, or destroy the asset? In some cases, one might infuse false data directly into the system (computer warfare) in an effort to confuse, frustrate, or deceive the enemy. In the case of Desert Storm, since the war was assessed as "taking some time" to accomplish, jamming was used as a cover to bomb defended sites, and A-10s (and other area munitions-carrying aircraft) were used to destroy undefended sites. The key to the plan was that the detection net was the first layer to be dealt with and neutralized.

Locating and Identifying as Information

Acquisition systems' primary responsibility is to locate and identify intruders. While they may perform this task fairly easily for themselves and the fire-control/interceptor systems they are directly tied to, the matter is greatly complicated when the data must be processed and transmitted via an elaborate net. Digital computer technology—while enhancing the speed and capacity of the acquisition radar—makes information denial an ideal strategy.

The modern IADS' acquisition systems enhance centralized command because they make this critical data immediately available to key decision makers. Modern information technology significantly shrinks the organizational distance and time between enemy agencies responsible for locating and identifying air intruders and key leaders responsible for making weapons allocation decisions.

The IADS net's ability to locate and identify the actors more rapidly and transmit the information quickly to the appropriate key decision makers and weapons systems provides the combined C^3/IADS net considerable flexibility. The enhanced exchange of location and identification data makes a distributed decision-making system unnecessary if the enemy command structure can secure the communication apparatus. Conversely—because location and identification data flows so fluidly in the digital computer-enhanced IADS net—the enemy command structure can exercise centralized command through several different communication paths and from different locations. In the process of enhancing command-and-control opportunities for enemy IADS decision makers, however, the enhanced IADS has insinuated key enemy commanders and political leaders into the US SEAD calculus because they have become a relevant—perhaps key—part of the IADS process. The technology of the information-sharing IADS makes data that was previously difficult for leadership to obtain immediately accessible, *thus making the decision maker a valid—and vulnerable—target.*

The timely availability of location and identification data to the enemy command structure of the future IADS will tend to involve enemy leadership more and more in the weapons allocation process. This may have the effect of making the IADS process more reliant on direction "from the top." A central "big picture," refreshed by constant new location and identification data, is much more efficient for allocating weapons systems. Therefore, the "big picture" has replaced the limited autonomous capabilities represented by cumbersome intermediary structures such as filter centers or midlevel command-and-control facilities that characterized older analog computer-aided location and identification systems. It also induces leadership into a more direct command-and-control role because the direct links to the total IADS become more tangible. Therefore, the centralized decision-making process enabled by the information-sharing IADS has the odd effect of making the enemy leadership itself a prime SEAD target, especially for those IADS which are—or may become—dependent on direct command and control from key military and political leaders.

Location and identification are information-intensive activities. They are directly related to the capacity of the systems to process and transmit the data. The radars that perform these functions are increasingly indistinguishable as separate detection, location/identification, or tracking radars. No matter how sophisticated the communications net, the data itself is vulnerable. No matter what type of radar the IADS employs, the data it would transmit to the net would be critical to the efficient operation of the enemy IADS. Denying access to transmission of the acquisition data through jamming, deception, or using special forces to physically sever ground and microwave links or alter the data's characteristics within the computer net will severely degrade the IADS' overall ability to conduct air defense.

Information Denial in the "Endgame"

The "endgame" is an appropriate phrase to frame the culminating events that occur as a result of enemy IADS tracking, air defense weapon employment, and friendly aircraft actions to avoid being shot down. No matter what the outcome, there will be an end result in the process. Denying information to enemy weapons systems at this point of the

IADS process, while increasingly difficult, can still be accomplished. Information denial as a SEAD strategy expands the opportunities for friendly aircraft survival in the endgame.

Allocation of the enemy weapon system to destroy the intruder is the final contribution of the enemy decision maker, assuming the required data reaches the decision maker. Once the enemy decision maker relays the pertinent data and the decision to engage to the SAM site, the AAA site, or the fighter-interceptor base, and the intruder comes into weapons range, the final engagement sequence ensues.

In a sense, the endgame portion of the IADS process has changed little since Vietnam. The character of the weapons themselves may have changed, but the context pitting individual flyer against individual air defense system remains the same. Using information denial as the basis of SEAD strategy, however, opens a broader array of options to the American aircraft finding itself in this situation.

Information denial in the endgame takes many forms. Self-protection jammers can deny critical tracking and fusing data to enemy fire-control radars or projectile fuses. Some forms of jamming and deception can prevent the launch of a missile or the firing of some AAA systems. Specific communications jamming can sever the links between fire-control and command units or confuse the situation until it is too late to bring air defense weapons to bear. Physical removal of the radar or critical communication link via antiradiation missile or bomb is also a form—albeit very direct—of information denial.

The most effective employment of SEAD against the IADS to ensure a favorable outcome in the endgame, however, is an integrated attack on all levels of the IADS process simultaneously. A resort to "last-minute" information denial applied piecemeal against a fully coordinated IADS will—as the enhanced IADS comes to the fore—increasingly yield unfavorable results for the US attacker.

The reinforcing effects of information denial applied simultaneously to the entire IADS information process would cascade through the IADS and compound its problems exponentially as it attempted to recover. An unwitting outcome of the SEAD campaign against the Iraqi IADS during Desert Storm was the discovery of the truly synergistic effects of simultaneous attack. Not only were the individual elements of the Iraqi IADS paralyzed, they were paralyzed in a way which made system recovery almost impossible *because there was no way for Iraqi IADS decision makers to determine what had happened to them.* In systemic terms, the simultaneous application of SEAD against all levels of the IADS process neutralized the Iraqi internal feedback mechanism.

The entire array of the future IADS—from the most remote early-warning radar to the point-defense AAA weapons used to defend the command center—will be tied together. By breaking the links that transmit data, denying the data itself from detection, deceiving or confusing the enemy, or destroying key sites as part of a sequential campaign, friendly forces can overcome the threat. To successfully challenge the future IADS, friendly JSEAD must likewise be integrated to counter the rapidly evolving enemy IADS. JSEAD tacticians, strategists, and key decision makers must define their objectives in unambiguous terms to which all can agree and in terms where each objective is translatable into a clear pattern of action.

Lt Gen Charles Horner's observations on forming "the basis of what happened during operations Desert Shield and Desert Storm" serve as an excellent summary. Of the five areas that formed the basis, two were directly concerned with classic SEAD. If we include the expanded definition of SEAD used in this study, then four of the five are included.[2] The areas discussed were:

1. "Unity of command and the joint forces air component commander (JFACC)." These are a prerequisite for future JSEAD.

2. "The significance of strategic strikes at the heart of the enemy's governmental and command and control structure." Under the expanded definition of JSEAD, both the leadership and C^3 structures are valid targets for JSEAD of the future.

3. "The absolute necessity of suppression of enemy air defenses." This is a classic acknowledgment of the value of SEAD. However, as minimum friendly attrition becomes more and more a political goal of American participation in combat, the emphasis on SEAD may grow beyond its purely military contribution.

4. "Increased survivability through timely use of electronic combat." This is also within the classic realm of SEAD, as SEAD is part of the Air Force's definition of electronic combat.

5. "A logistic train to meet the needs of fluid and dynamic deployment and employment." This is not related to SEAD or JSEAD per se; it is related to waging war in general.

General Horner's observations are all the more remarkable when compared to the Vietnam, Bekaa Valley, and Libya era operations. In former wars, battlefield commanders considered SEAD and electronic combat as subsidiary concerns. As the IADS has evolved, so has tactical concern. The elements of combat embodied in JSEAD have become principal concerns. Just as air deniability posed a poignant challenge to American air superiority in Vietnam, the digital computer and the information revolution will pose a challenge to American air superiority tomorrow.

Current trends indicate that because air power was such an effective tool in Desert Storm, potential enemies have redoubled their efforts in reconstituting an even more formidable IADS. This increases the emphasis we must place on JSEAD. Success in combat may well depend on the JSEAD strategist's ability to decode the enemy's information-based IADs and translate the results into usable tactics.

Adjusting to Fiscal Reality

Economic and political undercurrents pervade the entire JSEAD strategy and tactics discussion. These undercurrents represent forces over which the military has increasingly less influence. They prescribe a contradictory set of "facts of life" that *constrain* and *enable* the growth of JSEAD strategy and tactics.

Of the many facts of life with which the military will have to deal in the future, three are especially pertinent to JSEAD's evolution. These three realities are intertwined in such a way that they cannot be analyzed in isolation.

First, US forces are not likely to acquire the advanced, highly expensive technology needed to defeat a highly evolved IADS threat given the current political and economic environment—

no matter how strong a case the SEAD community makes. The US domestic scene is preoccupied with fiscal constraint, the first of many negative effects that will impact the military. The political climate is focused away from the military; even as Desert Storm was raging, the military budget was being pared and major "needed" weapons systems were cancelled or discontinued. USAF Brig Gen William Campbell noted in the *Journal of Electronic Defense* that the

> generational advantage [enjoyed in Desert Storm] can quickly evaporate. If we stand still, others will not only catch up but will pass us. Just to stay even we must invest in technology to counter the same high-tech systems that served us so well in the past war. To stay ahead requires continuous improvements.
>
> Ironically, the need for these investments in our future comes at a time when the size of the resource pie is shrinking and competition for RDA [research, development, acquisition] funds is becoming more fierce. There will clearly be less money to build all of the capabilities we truly need. As a result, the winners in the competition for dollars will be those potential solutions that are affordable, are supported by solid principle-based arguments and are demonstrably needed.[3]

Electronic combat weapons that are designed for SEAD are among the most expensive in the military. They must compete with all other facets of the military budget, and though money spent on SEAD systems is probably more cost-effective than most, equipment to support JSEAD cannot be acquired using the paradigms that motivated pre-Desert Storm acquisition strategies. The plain truth is that we must change our paradigm to adjust to fiscal reality. Countering the modern IADS threat will require a shift in more than resource allocation; it will require a shift in the way we look at the threat. Second, joint warfare is not just "a smart way" to wage war against the future IADS: *it is the only way.* It is fortunate that individual service doctrines are converging because the wartime requirement to break the modern IADS will require both joint resources and joint strategies. Curiously, the competition for scarce resources among the services should logically have the effect of crystallizing the identities of each and pulling their doctrines farther apart. As far as a nascent JSEAD is concerned, however, this seems not to be the case, and this will make the transition to functional JSEAD tactics much easier. The current acknowledgment of this fact and positive moves to make JSEAD a reality will also help shorten the "learning curve" when US forces are next called to battle. We can struggle with the issues now or struggle with them later, but we will struggle with them *jointly* as it is the only card we have left to play in successfully battling the future IADS.

Third, while only tangentially related to US JSEAD strategy, the issue of collective security through coalitions with other nations will become more and more important to military power application. The relative share of national power exerted by the US military will shrink, and its use on the international scene—without coalition assistance—will suffer some limitations. The twin forces of a shrinking US military and the growing sophistication of the rest of the world's military establishment make it plain that effective application of JSEAD strategy in the future may require the services of friendly nations in a coalition strategy.

European SEAD assets played a significant role in the Desert Storm effort. The United Kingdom's air-launched antiradiation missile (ALARM), for instance, was employed more than 100 different times. This missile, while performing many of the same roles of the HARM, is configured somewhat differently. It has, for instance, a loiter mode, in which the system uses a parachute to maximize time over the target area prior to acquiring, locking on, and tracking to the target radar.[4] The ALARM's complementary use in overall *coalition* SEAD strategy represents an evolutionary step beyond JSEAD. The development of coalition SEAD strategy will be propelled by international facts of life—not by US military strategy. Jointness in this respect may be just the beginning of more generalized forces galvanized from a coalition arrangement which may employ the skills associated with the roles each of the services plays, but little of their organization or doctrine.

The passage from old SEAD to new SEAD, from shooter-based to information-based IADS, and from a certain to less certain world is fraught with ambiguity and danger. As we resolve the uncertainties associated with the emerging world, the paths we should choose toward developing new SEAD strategies will also become clearer. In sharpening the focus of the information-denial JSEAD strategy, there is an implicit danger. While the strategy may become crystal clear, it is meaningless unless we have created viable tactics with which to execute it.

SEAD Tactics: Variations, Combinations, and Innovations

Tactics are tools: old ones are as acceptable as new ones, providing they can do the job. One changes the tools when the environment can no longer be altered by the tools one possesses. In this respect, the adage is only partly true that if the only tool one possesses is a hammer, then all one's problems will tend to be seen as nails.

The tactics formulation process is first and foremost an exercise in pragmatism. A realistic appraisal of the environment and the setting of achievable objectives is the first order of business. SEAD planners must determine what can and cannot be done with the tools at hand. Next, they must decide how best to meet the objectives with available tools or acquire tools that will accomplish the task. Each tool has absolute limits, but using the tools in different combinations may reveal hidden synergies.

Every conflict since Vietnam has yielded new SEAD tactics based on this pragmatic process. Compared to the SEAD strategist, the SEAD tactician defines the war-fighting environment differently. As part of tactics development inherent to the process, the SEAD tactician evaluates the validity of general aerial warfare strategy as outlined by the air tasking order—not by overall battlefield objectives stated in more general terms by strategists. Hence, the relationship between SEAD strategy and SEAD tactics is different than for many other facets of war fighting. Because SEAD tactics must respond quickly to the rapidly changing dynamics of the enemy IADS environment to be successful, SEAD strategy must ultimately be responsive to SEAD tactics. Ironically, this tends to make SEAD strategy tactics-driven.

SEAD strategy began as an amalgamation of various tactics which proved successful in combat or exercises. SEAD strategists analyzed successful SEAD tactics and drew from them organizing concepts that bound the tactics together.

From this analysis, modern SEAD strategies were developed. This developmental process differs fundamentally from classic strategy formulation that begins with the big picture and breaks it down into smaller and smaller components. The reason SEAD strategy does not follow the "big-to-little" strategy formulation process is that traditional strategy formulation does not account for rapid technological change or the dynamic nature of the battlefield with which SEAD deals.

Paradoxically, successful SEAD strategy seems to have stemmed more from its close relationship with successful SEAD tactics than it has from a relationship with general military strategy. The contradiction this implies is staggering. On one hand is the requirement to match strategy with overall objectives and build tactics to support the strategy. On the other hand, historical evidence supports the fact that SEAD tactics traditionally have driven SEAD strategy. This situation must be resolved by creating SEAD tactics that can be translated in terms of SEAD strategies generated by overall objectives. Further, SEAD tactics must be formulated such that their effects can be measured in terms of SEAD strategy. Tying the information-based IADS into an overall military scheme in a meaningful way aids in resolving this contradiction.

Dwindling SEAD resources and the changing nature of the IADS threat force a reevaluation of SEAD tactics. Nonetheless, many tactics used in the past provide a solid foundation for developing new SEAD tactics with which to challenge the formidable modern IADS. Other new tactics will be agglomerations of older proven tactics developed to capitalize on weaknesses perceived in the evolving IADS. Some SEAD tactics will be truly new, having evolved from an entirely different vantage of the military environment. These tactics will require an array of advanced weapons that the SEAD community must be able to justify in order to acquire. The following suggested SEAD tactics are examples of an approach that seeks to combine elements of SEAD's evolutionary nature, modifies existing tactics to new circumstances, and introduces some new tactics based on information-denial strategy.

Variations: The "IADS Sweep"

The IADS sweep is a variation of the MiG sweep of Korean and Vietnam War-era vintage. The primary threat of the MiG sweep was the enemy interceptor. The MiG sweep is classic Douhet theory in that the only real threat to the airplane is another airplane, but it goes beyond that. The MiG sweep recognized the potency of the enemy air defense structure as a threat to aerial superiority over the battlefield and in the pursuit of strategic objectives.

The MiG sweep embodies a sense of timing, place, and sequencing. Timing must be such that the maximum number of interceptors would be scrambled, creating a "target-rich" environment. The air battle must be fought in a place of the tactician's choosing lest the enemy gain the initiative. Lastly, the sweep must be sequenced such that the interceptors engage the MiG sweepers and not the friendly bombers following closely behind.

SEAD, in a sense, provides a natural bridge between air superiority doctrine and tactics-driven strategy. Douhet did not foresee the threat to air superiority stemming from advanced, integrated ground systems represented by SAMs and AAA; nor did he foresee the integrated structure that would tie interceptors, AAA, and SAMs together in a lethal, coherent network. SEAD is the "other side of the coin," the answer to the unforeseen challenge to air superiority portrayed in the modern IADS.

The objective of air superiority supersedes who accomplishes it or how it is to be accomplished (i.e., with fighters). The objective in air warfare is to reign supreme over the battlefields and strike at the enemy's strategic center of gravity with impunity. A precursor to this is neutralizing the enemy's ability to defend against air attack.

The IADS sweep combines elements of the MiG sweep and offensive uses of SEAD to achieve air superiority. The sweep consists of fighters, jamming aircraft, antiradiation missile shooters, stealth attack aircraft, attack helicopters, unmanned aerial vehicles (UAVs), long-range artillery (MLRS, A-TACMS, etc.), and very accurate cruise missiles netted together with a C^3 structure, warning net, and real-time intelligence. One

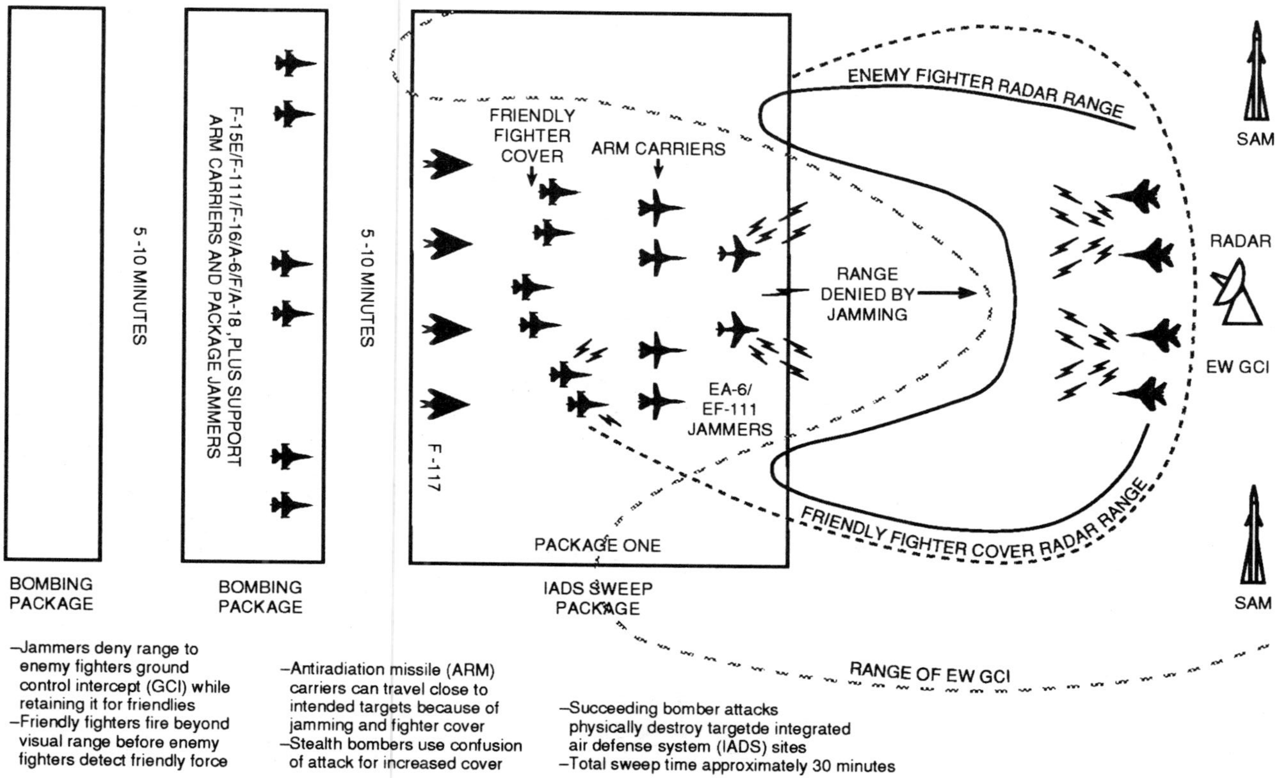

Figure 25. The integrated air defense system sweep.

might argue that the opening Desert Storm strike was, in reality, an unwitting IADS sweep. Figure 25 depicts the concept.

Note that the EF-111/EA-6 would be *in front* of the fighters. The reasons for this are twofold. First, the placement of the jammers would project the jamming as far forward as possible without interfering with other aircraft. Second, the jamming aircraft would blind enemy fighters while allowing friendly fighters to detect and track the enemy interceptors. The modern interceptor threat of the future will likely have avionics, range, and firepower equivalent to the United States'. The ability to deny the enemy interceptor absolute radar range in comparison to friendly fighters would give friendly fighters the ability to detect, identify, track, and fire on the enemy before the enemy even realizes where the friendlies were. Enemy ground controlled intercept radars would be jammed by the same aircraft that escorted the fighters. The jammers would obtain protection from the fighters placed immediately behind in a symbiotic relationship. The use of an IADS *sweep tactic,* to use the Army term, would build a designed kill zone entrapping the enemy fighters.

The jammers would also provide jamming protection against most of the acquisition radars, thus reducing the enemy's response time. HARM-firing aircraft (hopefully with capabilities similar to the F-4G) would target selected sites along the area or corridor to the general target set. If enemy radars did not radiate, they would be unable to supply the appropriate information to the fire-control system. The combination of jamming and ARM-firing would create another symbiotic relationship. The ARM-firing aircraft would be concealed from the radars by the jamming. This would achieve two tactical objectives. First, the jamming would be protective. The second would be that, since the enemy radar operator cannot detect the presence of an ARM launch, the site would continue to radiate, greatly enhancing the probability of the radar's destruction. This site's destruction would remove the danger the enemy fire-control system posed to the jamming aircraft and the follow-on friendly attack aircraft. Stealth aircraft would attack critical known sites of highest priority, usually command-and-control facilities. Immediately behind the IADS sweep—before the enemy could recover sufficiently to organize the air defense structure—would come the bulk of the attack force. The difference between the first night of Desert Storm and the depicted IADS sweep, however, is that the primary target of the IADS sweep would be the IADS net itself.

Unlike Desert Storm, the IADS sweep would be conducted sequentially by geographic area. While sequential in reality, the sweep's timing would appear to the enemy IADS as a simultaneous attack on all phases of the IADS information-processing network. Rollback of the enemy's IADS structure would be physical as well as psychological. The IADS structure would be jammed, disrupted, then destroyed piece by piece, element by element, with mutually reinforcing packages. The objective would be to wear away at the enemy and illustrate the enemy's powerlessness to the adversary's leadership. The knowledge that the enemy's entire country and defense structure were laid open to unopposed strategic bombardment with precision guided munitions would be a powerful inducement to come to terms quickly.

The IADS sweep brings to bear several principles of warfare: objective, concentration of forces, maneuverability, surprise, the use of the offensive, and economy of effort. The IADS sweep would strike at the heart of the enemy's defensive capability by using an overwhelming concentration of forces that would move through volumetric space and time to overcome local IADS defenses. The sweep would use surprise as a function of deception (EF-111/EA-6 jamming of enemy interceptor's radar after giving a "free look." The innate ability of aircraft to move rapidly as a cohesive element illustrates how an IADS sweep could maneuver rapidly throughout the enemy's territory as well as throughout the electromagnetic spectrum.

As an element of force designed to do something to the enemy, to seize the initiative, and to force enemy compliance to a prescribed set of behavior, the IADS sweep concept epitomizes offensive aerial warfare. Finally, by concentrating forces sequentially through the use of rollback tactics (reminiscent of the relationship between air power and sea/land power during the Pacific campaigns of World War II), we would make the most efficient use of our combat assets. The type of IADS American forces are likely to deal with in the future will possess a formidable array of long-range SAMs—a situation coalition forces in the Gulf did not face. The long-range SAM presents special problems for the IADS sweep and may even make the use of this tactic undesirable. This dictates the development of special tactics to deal with the modern long-range SAM.

Combining Old Tactics: SEAD and the Modern Long-Range SAM

The capabilities of newer long-range SAMs will exceed any single aircraft's ability to attack the missile with impunity, including traditional SEAD aircraft. The lethality of these systems, left unchallenged, is already extremely high and promises to get worse (depending, of course, on whose side it is deployed). If friendly forces encounter a modern long-range SAM system as part of a critical target set protected by the IADS, neutralizing it becomes a key task. To make matters worse, future acquisition systems for the long-range SAM may also include limited stealth-detecting capabilities.

Modern long-range SAMs have ambiguous radar signatures that make them difficult to detect or identify; thus, friendly jammers and fighters may come under attack without knowing they are being tracked. The modern long-range SAM is likely to possess a self-contained terminal homing mechanism that will make it autonomous of its parent radar once the missile has established lock-on. Its extremely high speed and maneuverability make it almost undetectable to the intended target.

An attack against the new-generation long-range SAM will require the use of multiple types of aircraft and other assets in the prosecution of this JSEAD mission. The JSEAD attack against the long-range SAM must employ a combination of tactics to include deception, brute-force jamming, antiradiation missiles, insertion of special forces in nonconventional warfare, and—as with any tactical operation—precise timing. Figure 26 diagrams the operation against the SAM.

The purpose of the jamming aircraft in suppressing the long-range SAM differs from the IADS sweep. The jamming in the SAM attack would be designed to place maximum jamming on a very small geographic area against very specific radars to

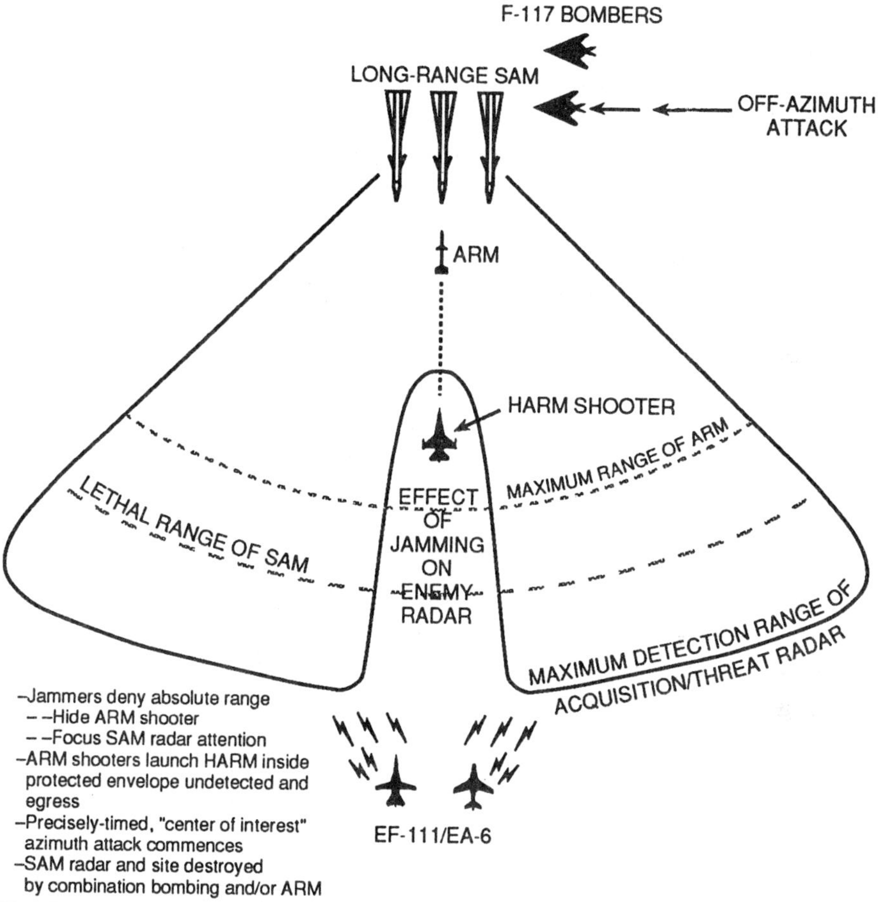

Figure 26. Attacking the modern long-range surface-to-air missile.

deny range information to the SAM's acquisition radar. This would reduce the absolute detection range of the radar on a specific azimuth, thus enhancing an antiradiation missile-carrying aircraft's ability to fire its missiles without becoming a target itself by "hiding" in an area of reduced radar range. Many long-range SAMs have lethal envelopes that could potentially place the antiradiation missile carrier at risk using normal stand-off antiradiation missile profiles. The jamming would be designed to shrink the acquisition/target tracking radar's detection range along a selected azimuth to allow the antiradiation missile-carrying aircraft to launch the missile undetected. Additionally, since most modern SAMs present their detection data via a computer-processed imaging technique, it may be possible to jam the radar without the operator's being aware of the "hole" in the radar coverage.

Modern SAM acquisition radars generally rely on acquisition data from sources other than the multiple-mode radar that characterizes the modern long-range SAM. Even though modern long-range SAM target-trackers have the capability to perform both acquisition and tracking functions, the preferred method is to acquire the information from a related system designed specifically for that purpose. Therefore, it is conceivable that specific jamming programs, tactics, and procedures would be moderately effective against even the most sophisticated radars. The objective for the jamming aircraft is to be effective enough in placing the antiradiation missile carrier in a firing position without jeopardizing either aircraft.

The jamming aircraft and the antiradiation missile carriers would be only part of the attack on the long-range SAM. At the same time that the EF-111/EA-6 would be preoccupying the SAM's interest, stealth aircraft would approach the site from an entirely different azimuth. Timing would be critical to the success of this ruse. If the stealth aircraft arrived too early, the SAM radar might not be neutralized by the antiradiation missile and the stealth aircraft might come under attack itself. If the stealth aircraft arrived too late, the IADS might have sufficient time to recover and bring other weapons to bear. The antiradiation missile in the SAM attack should be in flight during the time the attack aircraft would be closing on the target. Detonation of the antiradiation missile's warhead should occur just prior to the aircraft's arriving at the target. By timing the attack to coincide—nearly—with the antiradiation missile's arrival, the attackers would take advantage of the confusion caused by the blast and the sudden removal of the acquisition/target-tracking radar.

There is no such thing as a perfect plan. This set piece is not meant as the presentation of a surefire way to kill SAM systems that outrange, outgun, and outperform US single-systems' ability to neutralize them. The presentation of the SAM attack combining old tactics in a new way serves to illustrate the synergistic effects that a SEAD team would achieve—even against a system that is technically superior.

Innovations: Information Denial via Computer Warfare

Information denial as the basis of SEAD strategy leads to the creation of an entirely new family of tactics. At the base of the modern IADS network is the information network supported by the digital computer. SEAD tacticians, aware of this fact, could develop tactics that attack the heart of the IADS structure itself: the digital computers that support the IADS network.

SEAD tactics developed from following this line of logic take on an almost science fiction aspect, yet clearly they are achievable within the bounds of current technology. Direct attack on the computer network could be conducted by attacking critical power-generation equipment or fuel supplies that support the power generation. Since the technologies that support the IADS network will likely be international in nature and share a common industrial base, resupply of used or damaged parts must necessarily come from these same international sources. It should be fairly easy to gain access to these sources in the interdependent international computer industrial infrastructure. By introducing a device with embedded software containing an undetectable computer virus, the IADS structure would collapse once the infected hardware was tied into the IADS network.

Another method of direct attack on the computer network takes advantage of the nature of modern semiconductor technology. In order to achieve rapid processing speed, the wafers making up the semiconductor must be very thin. This makes them vulnerable to a phenomenon known as electromagnetic pulsing. This phenomenon induces very high levels of power across the computer's microcircuits, effectively burning them out. Current technologies are capable of producing weapons that can focus electromagnetic pulses against selected enemy computer centers to render them useless.

A more indirect form of attacking the computer could focus on the microprocessor itself. Certain types of jamming are known to affect microcircuit processing by deceiving the software into passing false data to other portions of the computer network. By developing jammers and jamming techniques specifically for this purpose, a whole range of IADS networking from radar processing to data transfer could be neutralized.

The intent of suggesting attacks on the IADS computer network characterizes the type of innovation required to defeat the modern IADS. The innovative techniques described here are direct logical outcomes of information-denial strategy based on a realistic appraisal of the modern IADS's strengths and weaknesses.

Joint SEAD Strategy: Dividing the Turf

It is not likely—or desirable—that the SEAD functions each service performs on its own behalf be stripped away in the name of JSEAD. What is desirable is the identification of the most likely areas where each service can help the others and the areas where JSEAD as an overall binding concept will enhance the war-fighting capability of the entire force.

The AirLand Battle (or airland operations in the parlance of the AirLand Battle future) and the "land-sea interface" (a term appearing in several naval publications) provide examples of how JSEAD concepts can apply. The airland operations concepts are driven by the idea of the extended, nonlinear, fluid battlefield that is reachable by long-range Army/Air Force systems—both defensively and offensively. The area beyond the fire support coordination line has traditionally been the responsibility of the Air Force (see fig. 27). However, the acquisition of new systems has given the Army an air-delivered power-projection capability that has radically altered the way the Army perceives the battlefield. Since Army and Air Force aircraft and long-range missiles will operate in this fluid area, the boundary line that formerly provided a relatively clear division of turf is not as distinct. The division of turf now appears more defined by weapons ranges, lethality, and the ability to fix an enemy force to create a momentary combined air/ground kill zone. Creating this kill zone seems

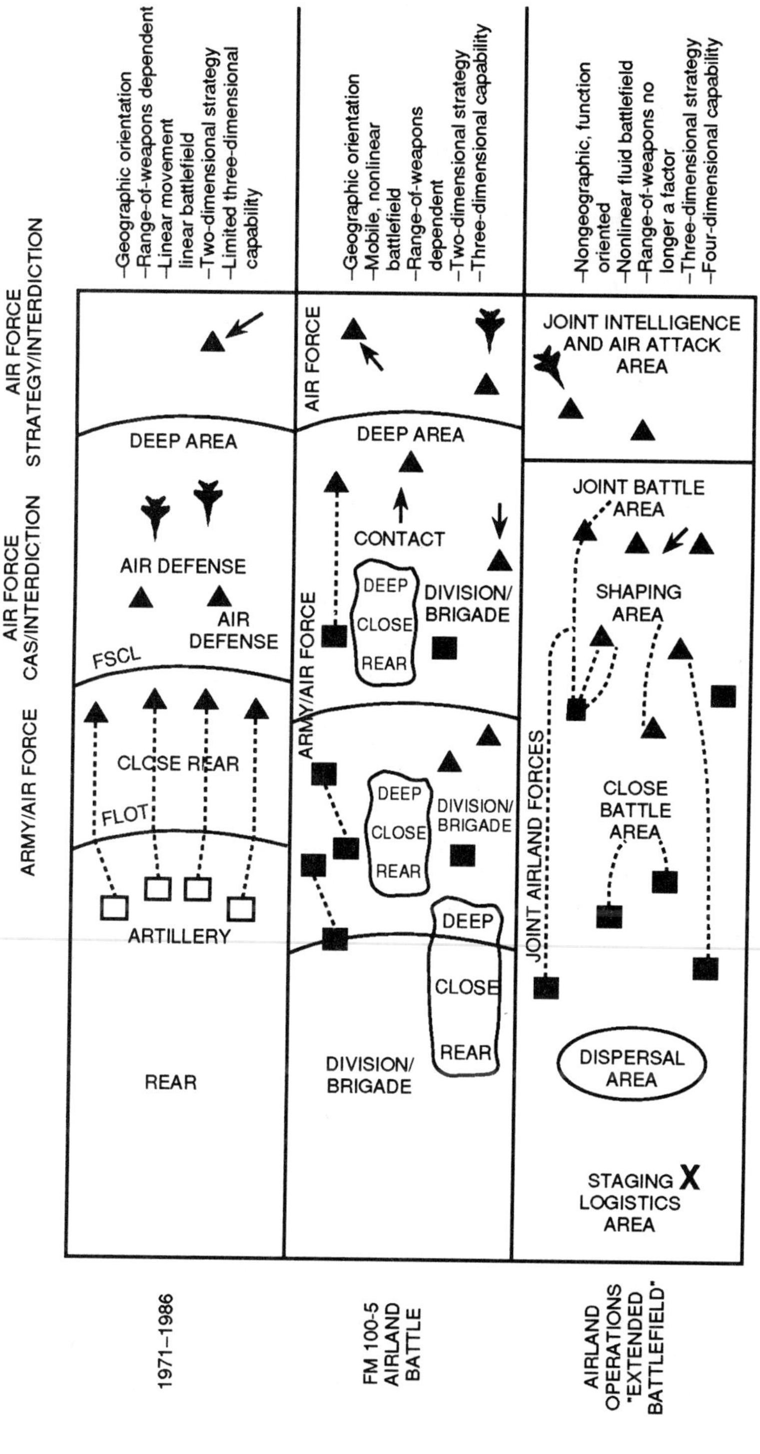

Figure 27. Evolution of Army/Air Force SEAD turf.

dependent on bringing as much firepower together as rapidly as possible to destroy the enemy enclave.

There is irony in the new Air Force and Army concepts of the battlefield. The new Army concept has strong overtones of flexibility, speed, range, lethality, and precision—which are Air Force concepts found in the 1992 AFM 1-1, the Air Force's primary doctrine document. This Air Force concept—with respect to ground operations—has strong overtones of initiative, agility, synchronization, and depth—which are found in Field Manual 100-5, *Operations,* the Army's primary doctrine document.

It is important for the Army to deny the enemy the use of the air space over the battlefield to protect ground forces. It is important for the Air Force to destroy the enemy's defensive capability against air strike so the Air Force can control the air space over the battlefield. Therefore, it is important to both the Army and the Air Force to cooperate in the destruction of the portion of the enemy IADS that affects both on the fluid battlefield. Figure 27 shows the evolution of the cooperative measures in SEAD (JSEAD, as defined by the Army) over time and provides a speculative view of how the turf should be divided to meet the future threat. Notice that the essential features that have changed are (1) the range and accuracy of the weapons, (2) the requirement for precise, timely intelligence, (3) the priority assigned to SEAD support by the Army, and (4) who controls the SEAD assets.

The range of the newer Army systems places them in an arena with which the Air Force has been familiar for at least two decades. The types of mobile firepower executed essentially through air power in support of ground operations can benefit from an understanding of air power strategies, especially where SEAD is concerned. The Army's air power projection capability on the nonlinear battlefield depends on how well the enemy's battlefield air defense is suppressed. Air Force SEAD concepts, tactics, and strategies seem an excellent starting point for greater success in combat in these areas.

The requirement for precise, timely intelligence is not limited to the new airland operations concept, though its effects will be much greater on the extended battlefield. The paradox of the "precision-guided war" is that it rests on precise, timely, relevant intelligence. The precision-guided munition is only as precise as the intelligence concerning the target.

JSEAD's intelligence requirements are especially vulnerable in the dynamic war against future air defense structures. The speed with which radars turn on and off, the speed with which they relocate, the rapidly changing frequencies and waveforms enemy radars emit, the rapid rerouting of entire communication paths—all require near-real-time (nominally five-minute-old data) intelligence updates at the very least. While most attacking aircrafts' intelligence requirements can be met by near-real-time information, radar jammers and Weasel-like aircraft will require the data described above *on demand.* When they need it, they will need it immediately.

The presumption that Army/Air Force assets will be able to maintain the initiative and that part of that initiative will be to drive the battle at a pace the enemy cannot adapt to is driven directly by knowing what center of gravity to hit next. The ability to acquire and distribute timely, relevant intelligence becomes critical to the JSEAD effort in airland operations, no matter how the turf is divided.

The priority assigned to JSEAD support has gradually increased as a function of the stake the Army has invested in air operations. JSEAD appears to have reached a threshold that forces traditional concepts of fire support to change. As air power becomes more critical to Army operations, fire support requests for JSEAD may come to supersede all but life-threatening, battle-losing propositions.

The airland JSEAD venture is no different than any other joint venture in one respect: there can be only one commander. Assets identified for JSEAD must be controlled by one agency, allocated by one agency, and employed at the direction of one agency.

The joint forces air component commander currently controls the EC cell, which is the responsible agency for JSEAD. The composition of that cell is not the exclusive province of the Air Force; in fact, as JSEAD matures and the war-fighting concepts embedded within it become better understood, its composition and leadership will reflect its objective-oriented nature irrespective of service. There will be fewer and fewer turf battles in the airland operations JSEAD arena for the simple reason that the turf will be defined by function and objective more than by geography.

Operations in the "land-sea interface" pose some intriguing problems. First of all, how far does the air power portion extend over the land-sea interface? How far inland does the land-sea interface extend? Who has operational control over the assets and the battle that rages in the land-sea interface? The same general questions concerning division of the turf that plagued airland operations prevail in the land-sea interface.

Navy EA-6s, A-6s, F-4Gs, EF-llls, and F/A-18s operating in the narrow land-sea margins could have a devastating effect on the enemy's localized ability to detect, locate, identify, or track any friendly aircraft being projected from the sea. The 360-degree attack from the land and sea against the enemy coastal IADS would doubtless overload the information-processing network and confuse enemy decision makers preparing for a frontal sea assault. The rapid projection of helicopter-borne land forces to specific locations to seize key terrain or destroy certain critical nodes under JSEAD cover would greatly enhance seaborne power projection ashore.

JSEAD Tactics: So What?

The requirement to match abstract concept to concrete reality is especially poignant in the SEAD strategy-tactics debate because so much is at stake. The logic of the debate is straightforward. If a major portion of US military power becomes invested in power projection via air power, then SEAD will take on a proportionately greater meaning. If the modern IADS emerges as preeminent in the ceaseless struggle for air superiority over the battlefield, much of the American military prowess currently enjoyed will be at risk. If US forces go to war under such conditions, not only will many Americans die but the United States will have failed to exercise its national power to achieve desired outcomes. Because so much is at stake, the roles of SEAD tactician and strategist take on new meaning in this argument.

The SEAD tactician, as ultimate pragmatist, lives in a "black-or-white" world. JSEAD strategic concepts and holistic paradigms mean nothing if they cannot be translated into tactics that are tangible at the war- fighting level. If the tactics that derive from the JSEAD concepts do not pass this "gut-level" commonsense test, the effort will have been wasted from the tactician's point of view. The purpose of presenting the joint scenarios is to show that applying a "unity-of-effort" concept does make a significant difference in the way we plan for and fight wars—even at the relatively simplistic levels presented here. JSEAD strategies and tactics do make a difference.

The IADS sweep could never be developed using current paradigms because no one has the authority to mandate it, nor is there a battle-staff planning function capable of managing it. The attack against the modern long-range SAM with the weapons we will likely possess could not be accomplished without the assets of all the services. The turf wars of the past do not make sense in the long view.

The SEAD strategies and tactics described in this chapter are possible only under true joint war-fighting constructs where each of the services has pooled its resources for the common objective of neutralizing the enemy IADS. The strategies and tactics presented are not mission/role-driven; they are objective-driven. There is no quibbling over who does OCA or interdiction, SEAD or C^3CM, AirLand Battle or land-sea interface. The objective is to neutralize the enemy IADS with the resources at hand, to win one war with one commander, one chain of command, and one plan.

Notes

1. Hal Gershanoff, "EC in the Gulf War: An Exclusive *JED* Interview with the Director of Air Operations for Desert Storm [Maj Gen John Corder, USAF]," *Journal of Electronic Defense*, May 1991, 44.
2. Lt Gen Charles A. Horner, USAF, "The Air Campaign," *Military Review*, September 1991, 17–27.
3. Brig Gen William H. Campbell, USAF, "Electronic Defense on the Modern Battlefield," *Journal of Electronic Defense*, October 1991, 43.
4. Martin Streetly, "HARM and ALARM in the Gulf," *Jane's Defence Weekly*, 30 March 1991, 500–501.

Chapter 6

Where to Next: Conclusions and Recommendations

JSEAD, because it is increasingly becoming a leading edge of military power application in all services, represents a prescriptive model for the future. SEAD planners can hold the paradigms and strategies that are evolving in JSEAD up to other mission areas that are struggling with the transition from single-service applied doctrine to joint applications. One might be able to glean today from JSEAD what the shape of future war-fighting doctrine and strategy will look like five to 10 years from now.

As advanced as it is in some respects, JSEAD still falls short as a "true" joint war-fighting capability. Nonetheless, it is closer to this goal than many other joint endeavors. There are three principal reasons for this. First, technology has tended to impact electronic combat (and hence SEAD) before it affects many other war-fighting skills. Second, SEAD has tended to be a war-only thought construct; there have been virtually no peacetime or deterrence roles implied in its use. Third, air power is rapidly becoming the primary power-projection weapon of all services, not just the Air Force. JSEAD, as the key enabler of air power projection, would logically be the first of the war-fighting concepts to become more fully developed.

Conclusions

The suppression of enemy air defenses is the result of a logical evolutionary process whose engine continues to be technology. As technology has accelerated, so has SEAD. The nonlinear and asymmetric application of technology across the varied elements of war fighting has given SEAD an evolutionary boost when compared to many other elements of war fighting. Therefore, SEAD tactics and equipment may be distinctly unsuited for doctrines and strategies that do not reflect the same level of evolutionary development. The asymmetric growth of JSEAD may pose some integration problems in its future application. Its mode of operation may not coincide with the role given it by strategic and tactical concepts that belong to previous iterations.

SEAD as an Evolutionary Concept

SEAD has evolved from a purely defensive concept to one that encompasses a more offensive outlook. By analogy, SEAD has evolved from protecting a two-ship fleet of F-4s in Vietnam, as a primary role, to destroying the fabric of an entire Iraqi nation's ability to defend itself from air attack in Desert Storm. From single-ship "bombs on SAM-site" technology and tactics, SEAD now encompasses sophisticated jamming and antiradiation missiles that disrupt and destroy the SAM and its associated connections to the IADS without ever entering the air defense system's lethal envelope. It has evolved from single-site, one-on-one concepts of operation to integrated force application involving hundreds of aircraft, cruise missiles, helicopters, and artillery systems attacking an entire IADS.

Requirement for a Paradigm Shift

Threats to US interests clearly exist, but using the paradigms of the cold war to counter them has limited relevance. The international community's transition from "what it was" to "what it is going to be" provides an opportunity for creating a proactive, capability-based paradigm that encompassed mobility, flexibility, and cooperative peacetime military measures to ensure stability in areas of the world that threatened the peace.

The new paradigm needs to reflect the realities of the emerging world order. First, trends indicate that military power as an element of national power—any nation's power—will continue to diminish compared to its historic role. National power has been supplanted by economic and regional considerations. Second, as an instrument of national power, military power must have precisely defined objectives that are clearly translatable to military action. While military power as a whole will diminish as an influence on the development of international relations, *when it is used, it will be absolutely critical.*

Third, the availability of very sophisticated weaponry and integrated air defense structures to virtually any nation will have a leveling effect among potential adversaries as time goes on. The relative ability to coerce by threat of overwhelming force or technical capability—either as an aspect of deterrence or pressure—will diminish.

JSEAD, as the evolutionary successor to single-service SEAD, can provide a transitional model for proactive, capability-based war-fighting strategy and tactics. Because JSEAD has evolved from a reactive to a proactive strategy, its outlook is more suited to the uncertain threat environment of the future. JSEAD is flexible and mobile and can be tailored rapidly to meet the changing needs of the battlefield. There are many "tools" in the JSEAD inventory—both hardware and concepts that can be rapidly reroled, recombined, and retasked to perform their missions.

JSEAD is not just the integrated application of multiservice SEAD assets and concepts. The political and economic realities of the US domestic scene will probably diminish each of the armed forces' individual capability to project power below what it requires to accomplish its traditional, *individual* worldwide responsibilities. This will require more than a "smart way" to integrate assets; it will require some new strategic constructs that are couched in more relevant paradigms. Table 11 compares the old paradigm to the emerging new one.

Table 11

Comparison of Old to New SEAD Paradigm

OLD SEAD	NEW SEAD
– Employed as collection of individual service assets	– Pooled resources; single binding concept
– Resource rich	– Resource constrained
– Weapons driven	– Tactics driven
– Employment by geographic target	– Employment by functional set
– Iterative, sequential, geography oriented	– Nonlinear, nonsequential, objective oriented

The Lead/Lag Issue: Joint Doctrine and JSEAD

JSEAD can do more for the joint force commander than current doctrine allows because it is more technically and conceptually capable than current joint doctrine. The synergies created in Desert Storm by simultaneous jamming, launching of HARMs at specific and area targets, special operations applications against specific sites, MiG sweeps, and fighter protection zones formed the core of proto-JSEAD operations and literally paralyzed the Iraqi IADS/C^3 net. An extension of SEAD concepts—mostly Air Force SEAD concepts—into the four-dimensional shaping of the battlefield has profound implications for overall joint doctrine. JSEAD is growing faster than the doctrine that supports it, and the gap between the two is growing.

JSEAD concepts are moving rapidly away from defining objectives in terms of places or things. The evolving objectives of JSEAD are systemic. These objectives use a framework of time and three-dimensional maneuver across the entire battlefield. Sequential attack, while often desirable, is not a requirement. Depending on the particular case and the war objective set by the battlefield commander, JSEAD concepts may mandate tactics that fly in the face of convention.

The key is to recognize that JSEAD concepts follow holistic and radical approaches to strategy and tactics problems encountered on the battlefield. Both JSEAD practitioners and battle management staff must adjust their mind-sets accordingly.

Shifting the Focus of Criteria: Overall Objectives

The criteria by which SEAD is adjudged effective need to be changed, as do the ways SEAD effectiveness is measured. The engineering approaches to SEAD and the application of mathematical models no longer make sense when assessing the overall effectiveness of SEAD because SEAD is evolving faster than the means to assess it. JSEAD concepts have evolved to holistic concepts with specific application across a broad spectrum. The criteria that have measured SEAD's effectiveness have been denominated in numbers—numbers that indicate how many aircraft were saved or the probability of kill against a specific site. A criterion is needed that can measure SEAD's ability to degrade the total capability of the enemy to defend against an attack. Current approaches to criteria formulation and measurements of effectiveness cannot provide an insight into the best way to combine SEAD assets to achieve the maximum benefit of JSEAD, nor can they assess force-on-force effects that utilize different types of aircraft, special forces, artillery, helicopters, or munitions in different combinations against different aspects of the IADS. The current criteria do not assess the impacts of timing of SEAD application (e.g., sequential waves; random waves of moderate size; simultaneous, mass attacks with relatively large gaps of time in between; etc.).

An assessment of effectiveness demands a comparison of some sort, and the measure of effectiveness must be denominated in some understandable and preferably concrete way. SEAD effects have never been easy to measure in any concrete way. A comparison of a control set (without benefit of SEAD) to a variable set (with SEAD) is not a precise measure. Unfortunately, future criteria will probably require the addition of more subjective and interpretive measures to the empirical methods now employed. In short, the criteria need to be focused on relating overall objectives at every stage of the strategy-to-tactics continuum and using measures of effectiveness that can answer the statesmen's questions as well as the tactics practitioners'.

Recommendations

As technologically based and complex as JSEAD will become, it still serves the interests of winning wars. This means it must be demystified. The value of JSEAD is lost if it cannot be understood by the decision maker.

Education

Battlefield commanders need to know more about JSEAD. The best way to accomplish this is through exercises that force the electronic combat cell of the planning staff to spell out what they can and cannot do for the commander. This would expose the EC cell to the overall priorities and strategic decisions the battlefield commander faces.

The key, not surprisingly, is to increase the opportunities of information exchange and to exercise the system that coordinates JSEAD efforts. Increasing the numbers of exercises that include JSEAD as an integral part of their scenarios is clearly indicated.

The electronic combat community must educate battlefield commanders and their staffs on the utility of JSEAD in unambiguous language. This will require aggressive and positive action. The creation of a series of relevant briefings that could be shown worldwide on JSEAD capabilities would be helpful to tacticians and planners at the very highest levels. There is a spinoff effect that enhances the tactical thought processes of the entire community in such exchanges.

Training

JSEAD will require extensive cross-service training and knowledge. Currently, training is restricted to specific unit exposure. The wars of the future will demand a more thorough ability to work together immediately. It took six months to prepare the "multiple service applied SEAD" observed during Desert Storm. The six-month preparation time was a luxury not likely to be repeated.

When exercises such as Green Flag are offered, joint commands should require a mixture of JSEAD forces and staffs to work together. This will force the use of joint doctrine and require aircrews and staffs to build a joint plan rather than four separate plans dictated by current tasking procedures. This means, for instance, that Air Force SEAD assets will work with Army, Navy, and Marine SEAD assets to build a joint plan aimed at providing the best defense for friendly air assets while at the same time destroying the enemy's defensive capability—as part of a component exercise.

JSEAD should be part of all services' exercises in the future. If it is critical to battle success, then it is worth doing right the first time and worth allocating the resources to do it. The payback, as evidenced by Bekaa Valley, Libya, and Desert Storm, is worth it.

The move towards a multiple-role force structure has enormous implications on training. Most units will be required to perform multiple missions that call for additional training. The addition of a JSEAD-motivated training load demands simplicity and efficiency. The capability-based solution has already been applied to PACAF's F-16 SEAD assets with full knowledge that the HARM-firing system can provide greater capability than aircrews train for. To use that capability, however, would have required training at unsupportable levels. Therefore, PACAF made the decision to balance limited SEAD capability against training time for other missions and roles.

Likewise, JSEAD technology application must simplify the training process, not add to it. Any addition to most frontline fighting units' current training calendars will result in diminished overall readiness if new systems require an extensive amount of time to learn and use.

Equipment

There are several categories under the recommendations for equipment, and these are all tied to technology. The technology base itself tops the list, but it is—as a result of rapidly changing realities—difficult to distinguish where general technology research and development ends and military research and development begins. It is also difficult to distinguish between the sophisticated industrial base which supports SEAD technologies and the equipment the industrial base produces.

Three areas of SEAD technology are critically important. First is guaranteeing the health of the sophisticated technology base that supports the advanced research needed for the future. Second, the SEAD community needs to evaluate the long-range requirements of new SEAD equipment with respect to the information-based IADS. Third, the SEAD community needs to establish clear guidelines on requirements for maintaining what it already has for the short term. The SEAD community must develop plans for exploiting current technologies to their maximum extent before undertaking the expensive venture into new technological vistas.

Protecting the SEAD Technology Base. The technological base is under threat of extinction because there is no clear and present danger to justify its existence. The old paradigm that related a specific weapon system to a specific threat as justification for the weapon's procurement died and left nothing in its place. Clearly, the sophisticated information-based IADS of the future constitutes a threat—even if SEAD strategists are unable to identify who the threat is. This poses significant difficulties for the SEAD community. While it appears to be true that US SEAD forces will continue to maintain a sufficient edge over the sophisticated threat, these SEAD forces will not have the overwhelming technological superiority witnessed in Desert Storm. The only way to guarantee an overwhelming technological edge is to invest in technologies that exploit principal weaknesses of the potential enemy's IADS. The information-based IADS provides the focus for advanced SEAD weapons development.

Procuring Advanced SEAD Equipment. The key to electromagnetic radiation propogation and electronic equipment operation is electrical power. To jam future IADS radars or computer networks that link the IADS structure together will require new power-generation equipment capable of producing much more power than is currently available. This will enable the creation of a new family of SEAD weapons that can physically damage radar and computer microcircuit components from great ranges—including space—with extremely high levels of focused electromagnetic radiation. The Army, Navy, and Marines could develop high-power microwave weapons for ships and ground vehicles and extend firing range by mounting them on telescoping masts to extend their line-of-sight range.

Computer warfare offers a whole new vista of technology development. Software programs could be developed to be inserted via hardware substitution, or perhaps new electromagnetic radiation technologies might be created to imprint programs on critical enemy IADS information net components from a distance.

The increasing sophistication of radar technology available to potential enemies mandates newer radar detection equipment for US forces. This will include some ground and shipborne SEAD forces as well. As the enemy IADS expands to target weapons designed to neutralize it, these physical components of US JSEAD strategy will be exposed to enemy tracking radars of great complexity. US aircraft and SEAD-specific ground and shipborne units must have adequate warning to defend themselves.

Current radar-warning equipment is designed primarily to detect Soviet-designed radars and will have limited capability against the hybrid advanced IADS. It is likely that new technologies and radical engineering approaches will be needed to develop these systems. This will, unfortunately, necessitate long lead times and great expense.

New technologies guaranteeing that US forces maintain the overwhelming edge in future combat will not be cheap. If the new technologies are required to achieve air superiority over future battlefields, then the SEAD community must develop sufficient arguments to convince acquisition and procurement agencies.

New Acquisitions Using Current Technology. Current on-board capability of US aircraft to detect, locate, and identify enemy IADS equipment and bring weapons to bear has considerable room for growth. The technology to accomplish the tasks of locating most enemy emitters is still relatively inexpensive and is accessible over the short term (three to seven years). Because the general threat has diminished, we have some time to deliberate on how much and what type to acquire—as well as how much to spend.

For the Air Force, retaining the F-4G is a requirement until sufficient capability is built up in the F-16/F-15E fleet to accomplish this much-needed task. Without the Wild Weasel capability represented by the F-4G, friendly attacking aircraft would be severely limited in their ability to detect and destroy an emitter whose location was unknown.

For JSEAD in the long term, it means defining services-wide requirements to develop a means of disrupting and/or destroying pertinent components of the enemy IADS emitting array. For reasons imposed by fiscal constraint, JSEAD systems procurement agencies must seek areas of commonality. The advent of digital computer technology and high-speed electronic buses for aircraft gives US aircraft a flexibility hitherto undreamed of, and there is no reason why a good Navy idea cannot also become a good Air Force or Army idea.

As force levels shrink, the equipment we currently possess must become multiroled. Multicapable aircraft, admittedly, is not the most efficient for specific roles. But we no longer have the luxury of designating specific systems for specific roles.

The acquisition of easy and ready-to-use systems would reduce the preparation time for combat. Systems of this type are easier to store and maintain, have greater reliability, and, because they are ready to use sooner, provide the tactician with a greater range of options. The canisterized Army multiple rocket launch system and the Patriot missile are examples of this type of packaging. Future Air Force weapons systems could benefit from this type of packaging, especially advanced antiradiation missiles and air-launched (or ground-launched) radar harassment drones.

Under joint auspices, the SEAD community should pursue efforts to acquire a common radar harassment drone. The requirement for an electronic harassment system such as the defunct Tacit Rainbow still exists even though this particular system has fallen from favor. The synergistic effects of using both ARMs and harassment weapons on the battlefield would be devastating to the enemy IADS and would give the US war planner tremendous flexibility in choosing weapons and tactics.

Current technologies offer significant capabilities in relaying pertinent data on the dynamic modern battlefield to US combat aircraft. The SEAD community needs to pursue off-board sensing systems that can relay data to combat aircraft in near real time. There are technical and pragmatic problems with the approach, but for aircraft which do not need "on-demand" data and are not in a position to be jammed by enemy countermeasures, the concept is solid.

Current technology can also be galvanized to produce a much-needed precision location system. A system was designed to perform this function (the precision location strike system) and was cancelled after it failed to meet specifications in the early 1980s, but the requirement still exists for a system that can precisely locate a hostile emitter or site. A precision location system takes on new meaning in the age of precision guided munitions. If we are to reap the benefits of precision guided weaponry for SEAD as well as general targeting purposes, a system such as this is well worth acquiring.

Current technology also allows the SEAD community to provide more aircraft with an ARM-firing capability. In light of the Navy experience with their ARM-firing EA-6 radar-jamming aircraft in Desert Storm, the Air Force decision not to provide the EF-111 an ARM-firing capability (made in the late 1970s) deserves another look. Considering the SEAD role played by Army and Air Force helicopters during Desert Storm, JSEAD strategists might also want to consider the possibility of ARMing a select number of attack helicopters.

Organization. JSEAD structures will be limited by human resources and equipment. There will be less and less of both. The mandate to create structures where each service has access to the other's SEAD apparatus is clearly indicated. This applies to common equipment, battle-management procedures, tactics, and basic knowledge. Current organizational structures propagate procedures that are cumbersome, fraught with parochialism, and needlessly time-consuming. Speed is part of flexibility, and organizational inertia of this sort has an absolute limiting influence on JSEAD's ability to respond meaningfully in conflict.

Three adjustments should alleviate some of these shortfalls. First, the EC cell, which works for the JFACC on joint staffs, should be composed of members of all services by aircraft type, weapon type, and organization type. This may sound like an enormous staff, but it is actually quite small compared to the way all of the services currently organize themselves to conduct SEAD operations. The proposed EC cell/JSEAD staff structure appears in figure 28.

Second, there should be considerably more interservice exchanges in the area of JSEAD and electronic combat. As F-4G expertise disappears from the Air Force, an exchange of SEAD tactics and strategy concepts will become critical—especially for the Air Force. As we move toward a more capability-based US armed force structure, emphasis on tactics and specific weapons application against an enemy IADS will also become more important. The Air Force expertise for destructive, lethal SEAD resides with the Wild Weasel community. It is uncertain whether the corporate memory of other strategic and tactical communities could compensate for its loss. Therefore, the separate SEAD communities of each of the services need to pool their knowledge and expertise so that the tactical expertise gained over almost 30 years of using the Wild Weasel is not lost.

To this end, the individual services should authorize an increased number of JSEAD-based exchange tours, and the JCS should authorize those officers selected to be given

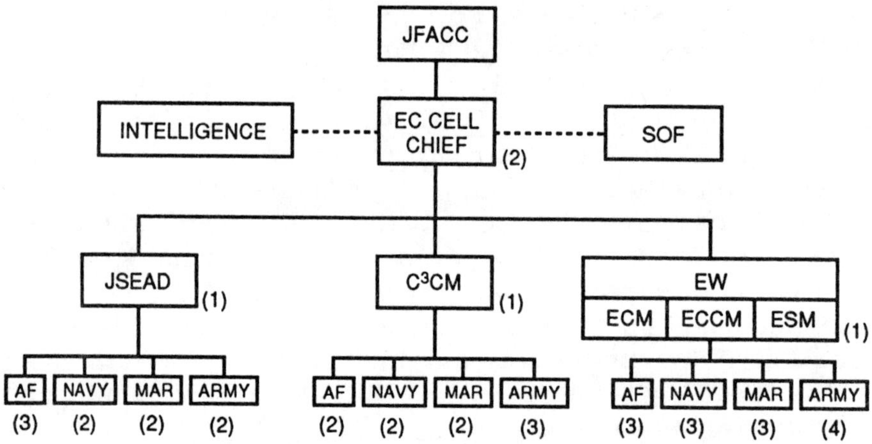

Abbreviations/Acronym Key

- C^3CM – Command, control, communications, and countermeasures
- EC – Electronic combat
- ECCM – Electronic counter-countermeasures
- ECM – Electronic countermeasures
- ESM – Electronic support measures
- EW – Electronic warfare
- JFACC – Joint force air component commander
- SOF – Special operations force

–JSEAD/C^3CM functions sometimes similar enough that same people can perform both functions; other times need separate staff.
–Individual functions manned by trained personnel, can be as many as 36 or as few as four. Ideal number in 8–12 range.
–Manning depends on situation, complexity of force structure, threat/IADS.
–Parentheses indicate prospective manning of full staff.

Figure 28. The JFACC's JSEAD organization.

joint-tour credit. This will enhance both the joint character of JSEAD and provide needed incentive in the electronic combat/SEAD-related career profiles.

Third, a JSEAD branch should be established at the joint staff J-3/J-2 level. This critical element is not currently addressed as a specific organizational function, and it ought to be. Sufficient expertise exists, but it is scattered and disorganized. It was fortuitous that the Air Force already had a proto-JSEAD staff in its EC cell during Desert Storm, otherwise those functions may never have been performed.

We cannot count indefinitely on American ingenuity and the apparent ability of the American warrior to create very effective ad hoc mechanisms when we need them. There is no reason why such an important function cannot be codified and institutionalized in a coherent manner.

Full Circle: The Strategy-Doctrine-Tactics Continuum

The overall direction of foreign policy is diffuse and wavering. There is no substantial guidance. The period of transition brought on by the demise of the Soviet Union presents an unparalleled opportunity for creating cohesive JSEAD strategies and tactics. The elements that belong to the natural evolutionary processes imbedded in SEAD make it ideal as a vehicle to enhance the change from single-service applied concepts to a true joint application of military power.

For the first time in a long while, service doctrines are converging. The realities of dwindling resources have forced us to review the basic questions: What are the legitimate functions of the military in an uncertain world? What are the limits of military power? How can the military—and SEAD—contribute more meaningfully? Do our methods and actions support the overall objectives of the system we are sworn to serve? For the brief while that the world may give us, now is the time to reflect on our purposes and the time to create visions of the future—a future we have a golden opportunity to influence, to build, and to protect.

Index